Front Cover: Marine Reservists (F) on the Ellipse behind the White House "Dress Right, Dress" in preparation for drill in 1918. (WarDept Photo 165-WW-598A-A, NatArch).

WOMEN MARINES IN WORLD WAR I

By

Captain Linda L. Hewitt, USMCR

HISTORY AND MUSEUMS DIVISION
HEADQUARTERS, U. S. MARINE CORPS
WASHINGTON, D. C.
1974

DEPARTMENT OF THE NAVY
HEADQUARTERS UNITED STATES MARINE CORPS
WASHINGTON, D. C. 20380

Foreword

The history of the first women to serve in the Marine Corps is a fascinating record of the dedication and drive of American women during World War I. The purpose of this monograph is to tell the story of the small band of women who answered the Corps' call for volunteers in 1918 with patriotism and enthusiasm. Long the object of interest and curiosity by modern-day Marines, the women Marines of World War I now have a lasting and fitting memorial.

The author, Captain Linda L. Hewitt, USMCR, is a native of the State of Washington and a graduate of Seattle University. She served on active duty from 1967 through 1974 and is presently a member of Senator Henry M. Jackson's staff.

In the interest of accuracy and objectivity, the History and Museums Division welcomes any comments, additions, and recommendations in regard to this monograph.

E. H. SIMMONS
Brigadier General, U. S. Marine Corps (Ret.)
Director or Marine Corps History and Museums

Reviewed and approved:
10 June 1974

Preface

A former Director of Women Marines, Colonel Jeanette I. Sustad, USMC (Ret.) originated the project of compiling data for a history of women Marines who served in World War I. In 1971, she asked various members of the Women Marines Association to interview surviving veterans throughout the country. A questionnaire designed to guide the interviewers as well as background information on the service of women Marines in the 1918-1919 period was developed by Lieutenant Colonel Pat Meid, USMCR. Lieutenant Colonel Meid, who authored the official history, Marine Corps Women's Reserve in World War II, originally published in 1964, accumulated considerable material on the earlier group of women Marines during her research. This was all made available to the author of this monograph.

The interviews conducted during 1971-1972, 29 in number, form a valuable archive of personal experiences of these pioneer women Marines. They have been used to supplement the official records which are sparse and elusive. Muster rolls of the time were checked exhaustively in compiling a roster of women who served, but it proved impossible to discover all the names making up the 305 women who were enlisted as Marine Corps Reserve (F). Much information was gleaned from contemporary magazine and newspaper articles, particularly from Leatherneck, Marine Corps Gazette, The Marine Magazine, Recruiter's Bulletin, and the New York and Washington daily newspapers. A small but useful collection of women Marine memorabilia, including photographs, letters, and clippings, was donated by various individuals as a result of publicity about the project.

In the course of my research, I was cordially received and assisted by staff members of the National Archives, the Library of Congress, and the Marine Corps' own History and Museums Division. I particularly would like to thank Mr. Ralph W. Donnelly of its Reference Section for his special assistance. The monograph was prepared under editorial direction of Mr. Henry I. Shaw, Jr., Chief Historian of the History and Museums Division. I am indebted to the present Director of Women Marines, Colonel Margaret A. Brewer, USMC, who continued and supported Colonel Sustad's original concept of the need for a history of women Marines of World War I.

Linda L. Hewitt
Captain
U. S. Marine Corps Reserve

Table of Contents

Introduction

Legend has it that the first woman Marine was Lucy Brewer who supposedly served, disguised as a man, on board the frigate Constitution in the war of 1812. (1) While there is no evidence that Miss Brewer ever wore a Marine uniform there can be no question about Opha Johnson, who on 13 August 1918 enrolled in the Marine Corps to become America's first woman Marine. (2) Her enlistment was a reflection of the dramatic changes in the status of women wrought by the entry of the United States into World War I.

The nation was already heavily committed to the support of the Allies when the declaration of war was signed in April 1917, and as thousands of young men rushed to volunteer for the Armed Services, and the draft gathered in hundreds more, the labor potential of women for the first time in the history of the United States became of monumental importance. In August 1917, four months after the Navy opened its doors to women in an effort to support the increasing administrative demands of the war, the Secretary of the Navy said: "In my opinion the importance of the part which our American women play in the successful prosecution of the war cannot be over-estimated." (3) In October of that same year the New Republic commented: "Our output of the necessities of war must increase at the same time that we must provide for the needs of the civil populations of the countries allied with us. Where are we to get the labor?...The chief potential resource at our command lies evidently in the increased employment of women." (4)

Overnight, organizations such as the National League for Women's Services and the Women's Committee of the Council of National Defense were established to coordinate and direct the activities of women across the country. Everyone, from housewives in Oklahoma to Park Avenue society girls was involved in the all-out effort to support the war. Much of the work fell within the area of volunteer labor and thousands of women were recruited to work at home or in local women's clubs making bandages, knitting garments, planting victory gardens, or canning a can for Uncle Sam. Women with more time moved out of their traditional roles and volunteered to hostess at canteens, organize food and clothing drives, or collect books and magazines for the boys overseas. Hundreds more canvassed the city streets promoting the sale of Liberty Bonds and War Savings Stamps or joined such organizations as the American Red Cross or the bloomer-clad Women's Land Army whose members helped farmers with the task of producing the Nation's food supply.

Private Opha Mae Johnson, the Marine Corps' first enlisted woman, shortly after her enrollment on 13 August 1918. (NatArch RG 127G Photo 515829).

The contribution made by the highly-motivated women's "volunteer" army was invaluable to the war effort; however, of greater historical significance was the breakthrough achieved by women in the gainful occupations. Contrary to predictions early in the war that women would never be put into "trousers or an unbecoming uniform and try to do something a man can do better," women donned the uniforms of elevator operators, streetcar conductors, postmen, and industrial workers, and ably carried on the Nation's business at home. (5)

Dramatic changes in the status of American women were to result from the wide-spread employment of women in United States industry. Although women were by no means new to the industrial setting, World War I was the first time in which they were employed in large numbers to perform skilled and semi-skilled labor. In aircraft plants, steel mills, and shipbuilding yards, hundreds of women were recurited to operate drill presses, lathes, millers, and other machines and hand tools requiring a high degree of dexterity. (6)

Although American industry and business were the first to feel the pinch of the manpower shortage, and the first to employ large numbers of women, by 1917, the military services were also faced with a serious personnel problem. Josephus Daniels, the Secretary of the Navy who was enthusiatic about the possibility of enrolling women in other than nursing billets, set his staff to work investigating the legal statutes regarding naval enlistments. While his assistants tried their best to find laws prohibiting female enlistments, there was in fact no legal basis on which to exclude women - naval law referred only to the enlistment of "persons." (7) So Secretary Daniels opened the Navy to women under the conditions of equal opportunity. In all, some 12,500 Yeomen (F), holding mostly clerical positions, responded to the nationwide "Call to Colors" broadcast in March of 1917. (8)

The Army which also desperately needed administrative support and would have welcomed women was not so fortunate. The law pertaining to the Army specifically called for the enlistment of "male persons." In spite of a deluge of requests from hundreds of commanders, all attempts failed to change the law and women were not to be enlisted into the Army until the outbreak of the Second World War.

The position in which the Marine Corps found itself with regard to personnel was similar to that existing in the other services. The way in which the Corps sought out to remedy its situation, with many colorful rememberances by the women who served, will unfold in the following pages of this monograph.

WOMEN MARINES IN WORLD WAR I

Authorization

By July 1918 the demands of the war hit an all-time high, the heavy fighting and mounting casualties abroad were increasing an already acute shortage of trained personnel and as fast as men could be spared, they were sent to join Marine units at the front in France. When it was discovered that there was a sizable number of battle-ready Marines still doing clerical work in the United States who were urgently needed overseas, the Corps turned in desperation to the female business world. Major General Commandant George Barnett, in an effort to determine how many men could be released, dispatched memorandums to the offices of the Quartermaster, Paymaster, and Adjutant and Inspector asking for an analysis by the directors of each as to the feasibility of using women as replacements for male troops. In every case it was agreed that there were areas in which women with clerical skills could be utilized on an immediate basis. Interestingly enough, although it was estimated that about 40 percent of the work at Headquarters could be performed equally well by women, it was believed that a larger number of women than men would be needed to do the same amount of work. The opinion expressed by experienced clerks was "that the ratio would be about three to two." (9) The highly competent performance of the women reservists throughout their participation in the war proved the error of this early opinion. In addition, the Major General Commandant was strongly advised that if women were to be enrolled "it would not be desirable to make the change suddenly but gradually" to allow sufficient time for each woman to be instructed by the clerks they would be relieving. On the basis of these conclusions and recommendations, General Barnett wrote a letter to the Secretary of the Navy on 2 August 1918 requesting authority "to enroll women in the Marine Corps Reserve for clerical duty at Headquarters Marine Corps and at other Marine Corps offices in the United States where their services might be utilized to replace men who may be qualified for active field service." (10) In a letter dated 8 August 1918, Secretary Josephus Daniels gave his official approval to the request and authority was granted to enroll women as members. (11)

Recruiting and Enrollment

Overnight the word was spread via newspapers and enthusiastic Marines and on 13 August 1918 women by the thousands flooded into recruiting offices across the country. Mrs. Opha Mae Johnson, who was working at Headquarters Marine Corps as a civil service employee, was enrolled on 13 August 1918 and holds the honor of being America's first woman Marine. Mrs. Johnson was assigned as a clerk in the office of the Quartermaster

A fanciful woman Marine was used by Howard Chandler Christie as his model in this World War I recruiting poster. (NatArch RG 127G Photo 518347).

General, Brigadier General Charles McCawley, and by the war's end was the senior enlisted woman with the rank of sergeant. (12) The Marine Corps had very definite ideas as to the type of women it wanted as members of the Corps. Recruiting officers were instructed to enroll only women of excellent character, neat appearance, and with business or office experience. While the greatest need was for stenographers, bookkeepers, accountants, and typists, "applicants who have had considerable experience in handling correspondence and general office work and can show evidence of exceptional ability in this line will be given consideration." (13)

Recruiting orders stipulated that women reservists were to be between 18 and 40 years of age but that an applicant slightly under 18 years of age, who is in every respect very desirable, may be enrolled with the consent of her parents." (14) In addition, each applicant was required to submit to a thorough physical examination before her final acceptance. While this seemed a simple task, it presented unusual problems as the enlistment requirements established by Naval Medical Regulations were designed for men only. Medical officers, it seems, were as uneasy with the situation as were the women applicants when it came time for the examination. In an effort to establish a policy guiding medical officers, Headquarters released a "circular" on 14 August 1918 giving detailed instructions as to the enlistment requirements expected.

While the Bureau of Medicine realized the obvious source of embarrassment caused by the examination, it was strongly felt that thoroughness should not be sacrificed because of it. Accordingly, each medical examiner was instructed to "use such tact and courtesy as will avoid offending in any way the sensibilities of the applicant. He shall not, however, by such attitude, allow himself to deviate from a proper fullfillment of all the requirements. The applicant should be previously instructed to arrange her clothing in a way that will insure ease, facility and thoroughness in the examination. A loose gown of light material will not interfere with the examination or taking of the measurements. Corsets should invariably be removed." (15)

The women selected for duty with the Marine Corps were enrolled as privates in the Marine Corps Reserve, Class 4, for a period of four years. Each applicant was required to furnish three letters of recommendation and if possible was given an interview with the head of the office in which she was to be assigned. In cases where applications were received from great distances away and it was apparent that the applicants involved possessed the required skills, Headquarters Marine Corps ordered that "they should be instructed to report at the nearest recruiting station for examination and a test as to ability and in such cases should be required to furnish additional letters from former employers of undoubted reliability." (16) Such was the case with Miss Sarah Jones who because of her excellent qualifications was ordered from Meridian, Mississippi to New Orleans, Louisiana for interview and processing.

Following her enrollment on 27 September 1918, Private Jones returned to her home in Meridian to await orders from Headquarters Marine Corps directing her to report. Within a matter of days the official orders arrived and with excited expectation she boarded the Pullman for the two-day train trip to Washington. Upon her arrival Private Jones was assigned as a secretary to the non-commissioned officer in charge of medals and decorations where she was to remain until the war's end. (17)

The enthusiasm with which women across the country responded to the call for volunteers was amazing. In New York City alone, 2,000 hopeful applicants lined up at the 23d Street recruiting office in reply to a newspaper article which read that the Marine Corps was looking for "intelligent young women." (18) Colonel Albert McLemore, the Officer in Charge of Marine Corps Recruiting, was on hand to ensure proper screening and processing of the women. Among the five ladies chosen on that occasion was Miss Florence Gertler who recalled the day's excitement:

> Male noncommissioned officers went up and down the line asking questions about experience, family responsibilities, etc., and by the process of elimination got the line down to a few hundred. Applicants were interviewed by one officer and finally given a stenographic test. Colonel McLemore conducted the shorthand test and dictated so fast, that one after another left the room. Those who remained were taken, one-by-one, into Colonel McLemore's office and told to read back their notes (I remember that I made a mistake on 'Judge Advocate General,' never having heard the word, I thought it was 'Attorney General.') If the colonel was satisified with our reading, we were required to type our notes and timed for speed and accuracy. More and more applicants dropped by the wayside, until only five of us were left. We were told to report back the next day for a physical examination. (19)

Within a week all five were called up to be sworn in and issued orders for duty at Headquarters Marine Corps in Washington D.C. Florence later learned that Colonel McLemore called them his "100% girls" because of the unusually high speed and accuracy requirements placed on them that first day of recruiting. Because of her experience as a stenographer and her superior test results, Florence was assigned as a secretary to the Assistant Adjutant and Inspector, Captain France C. Cushing. (20)

A flood of women applicants in the New York Recruiting Office greeted the announcement that women Marines were being enlisted. (WarDept Photo 165-WW-598A-12, NatArch).

Dressed in men's uniforms, (l to r) Violet Van Wagner, Marie Schlight, Florence Wiedinger, Isabelle Balfour, Janet Kurgan, Edith Barton, and Helen Dupont are sworn in as privates at New York. (NatArch RG 127G Photo 530552).

The Marine Corps was unwavering in its determination to accept nothing but the best, most highly trained women possible. Across the Nation the story was the same. Among the thousands of women who burned with partiotic desire to serve the country as a woman reservist only a select few were able to meet the rigid requirements demanded by the Corps. In all, only 305 were enrolled during the four months of recruiting. (21) It was not unusual to find that out of 400 applicants only one had been chosen or that out of 2,000 only five were found to be qualified. Time could not be wasted training. The women chosen were enrolled with the idea that they would be able to step in and be on their own within a matter of days and in some cases within a matter of hours after their arrival.

These women had to be "top notch" and had to be able to demonstrate their qualifications under the pressure of a stiff recruiting examination. One young woman by the name of Elizabeth Shoemaker was not to be discouraged in her effort to qualify for enrollment. Miss Shoemaker was working as a stenographer in New York City when she saw the notice in the New York Times that enlistments would open up the next day at the 23d Street recruiting office. Although she arrived early, the line was already clear around the block and down the street from the front of the station so she took her place at the end and began to inch toward the door and up the stairs to the office where interviews and examinations were being conducted. One by one the women were told to step aside "You had to be 100% perfect mentally and physically...and you also had to be a speed stenographer." In spite of the fact that she worked as a speed stenographer, Miss Shoemaker failed the typing test that day and was told that she could not be used:

> I was terribly disappointed, of course... the next day I went back and for no good reason... I was only 17, I hadn't had my 18th birthday yet... I did my hair differently and had dressed differently. I thought I might be recognized, and I was. This amazing colonel said, "Weren't you here yesterday"?... I remember him hesitating and when I said "Yes, I was" he got up and leaned over the desk and shook my hand and said, "That's the spirit that will lick the Germans, I will allow you to take the test again"! (22)

This time she passed the test with flying colors and was on her way to Washington D.C. for duty where she was assigned as the secretary to the chief clerk in the office of the Adjutant and Inspector.

The spirit of these first women Marines is indeed something to be admired. While the reasons behind their desire to enroll in the Marine Corps were as varied as their numbers, the patriotic duty felt by each to be of service to their country in time of conflict was the force stronger than any

Two hopeful applicants to become women Marines are interviewed by a recruiting officer in Boston. (War Dept Photo 165-WW-598A-7, NatArch).

Six newly enlisted privates from Boston pose for their picture before taking the train to Washington, D. C. (WarDept Photo 165-WW-598A-1, NatArch).

other which accounted for the overwhelming response to the call for volunteers. Some, like Private Theresa Lake, who was working as a statistician for the Texas Oil Company, has a special reason for enrolling--her sweetheart had given his life for his country. Others, like Private Lillian Patterson, served to support her husband at the front and Private Mary English her brother. (23) The women of 1918 wanted to be totally involved, and, recalled Private Mary Sharkey, the opportunity to be members of "such an elite branch of the service" was the opportunity to serve in the most meaningful way possible. (24) One young woman was so filled with the desire to be a Marine that after being told that she did not meet the requirements because she was not a stenographer burst into tears, "But I'll do anything," she pleaded, "anything." (25)

The Macias family of Jersey City, N.J. was well represented in the Corps when its only daughters, Edith and Sarah, were sworn in in New York on 5 September 1918. Edith and Sara had been working as stenographers in New York when they saw the classified advertisement that the Marine Corps was accepting women. They both had good clerical and stenographic experience and were sure that they possessed the necessary qualifications for enrollment. Sarah was 27 years old, and Edith, although only 17, had been telling her bosses and the D. Appleton Publishing Company that she was 19, so, she thought, why not the Marine Corps too?

After a lengthly discussion with the recruiting officers concerning background and experience, the girls were given a physical examination, medical shots, and the stenographic test. Edith in her excitement failed the test and was going to be turned down, so Sarah bravely asked permission to talk to Colonel McLemore, the senior recruiting officer, and succeeded in convincing him of Edith's ability. They were both sworn in on that day and ordered to report the following Monday to Washington for assignment. Although their parents "nearly fainted" when the girls broke the news of their enrollment, the two silk stars found hanging in the window of the Macias home the next day was proof indeed of the tremendous pride felt by their family. Privates Edith and Sarah were detailed for assignment together as stenographers in the office of the Adjutant and Inspector. Upon their discharge, Edith had attained the rank of private first class and Sarah the rank of corporal. Another pair of sisters who patriotically enrolled in the Corps together was Helen Constance Dupont, whose husband was serving in France, and Janet Kurgan.

The newly enrolled Marine Reservists (F) were as proud of the traditions of the Corps as the men themselves. While some of their families and friends might have reacted with surprise and in a few cases shocked disbelief, the women themselves were delighted with their achievement. In a letter, one of a series published in The Recruiters' Bulletin, Private Martha Wilchinski ostenisibly broke the news to her sweetheart fighting in France:

Dear Bill,

I've got the greatest news! No, I haven't thrown you over; I'm still strong for you, Bill. No, it's no use; don't try to guess. You're not used to that much mental effort, and you might get brain fag. Besides, you'd never guess, anyway. Now, listen and try to get this. I know it'll be hard at first, but it'll grow on you after a while. Are you ready? Well, then,-- I'm a lady leatherneck; I'm the last word in Hun hunters; I'm a real, live, honest-to-goodness Marine! The process was painful I admit, and lasted for thirty-six hours, but I survived it all right. Our future together doesn't look so black to me now. Don't be surprised if you see me mentioned for a Croix de Guerre or something. You know me! I'm not looking for sympathy or anything, but honest, I've been through an awful lot. They've done everything to me except punch my name out on my chest. That's coming soon, I guess. But I'll begin from the beginning and tell you everything ad seriatim. That's Latin. It means, 'Go to it, kid.' You know I always had a kind of a hunch that the Marines would realize the necessity of women some day, so I was laying low and waiting. Well, when I heard they had at last hung out a sign at the recruiting station -- 'Women wanted for the United States Marine Corps' -- I was ready. 'Mother,' says I, 'give me your blessing, I'm going to be one of the first to enlist.' I was there when the doors opened in the morning. I was one of the first all right -- the first six hundred! You'd think they were selling sugar or something. Well, when the crowd heard that you had to be willing to go anywhere as ordered and you had to be a cracker-jack stenog, they thinned out some. And from what was left the lieutenant picked out twelve to go over to the colonel and have him give us a double O. I was one of them, of course. I'm not looking for applause, but you know I always said with colors when they change every season. The colors don't change, of course, but some smart guy wants to make it hard for everybody else so he calls green, chartreuse, and yellow, maize. Then they took my finger-prints. They'll know me when they see me again. The nurse couldn't find any marks on me for purpose of identification. "Take a picture of my freckles, nurse," said I. Some kidder, eh? And, I've got a terrible confession to make to you. You know what I said, no secrets between us. They took my height in my stocking feet. It wasn't fair; nobody had ever done that to me before and

Marine Corps artist Morgan Dennis pictures a newly arrived woman Marine waving "Good-bye" to the Marine she released for overseas duty. (NatArch RG 127G Photo 528855).

I told the doctor as much. I'm a terrible shrimp, and I don't know whether you'll want me when I tell you. I'm sixty-two inches. Isn't it heart-breaking? I felt as big as a yardstick when I heard it. But you know me, Bill, I'm a sport. You can always have your ring back. There's still nine installments to pay on it, anyway.

Well, only three of us came out alive. The others had fallen by the wayside. Then the colonel came in and told us to come over and be sworn in. I'm going to tell you something. I'm not bragging, but it isn't every private that's sworn in by a colonel. It was terribly impressive. Something kept sticking in my throat all the time. I don't know whether it was my heart or my liver. I had to swallow it several times before I could say, 'I do.' Then they took a movie of us. I'm not throwing 'bokays' at my self, but you've got to admit it, the kid's clever!

And then I got my orders. Travel orders they call them. But that's only to make it hard. The only traveling I have to do is to come down from the Bronx in the new subway. I'm so worried about those orders, I sleep with them under my pillow at night and wear them around my neck during the day.

I got some good tips from the boys. They said if you want to scare the captain just click your heels at him. I don't remember whether they said click or kick; I guess they meant kick. And another thing they said. When I'm made a sergeant, I mustn't stand for being called 'Sarge.' Nothing doing on that 'Sarge' stuff. They'll have to call me anything that's in the Manual. I hear some people are giving us nicknames. Isn't it funny the minute a girl becomes a regular fellow somebody always tries to queer it by calling her something else? There are a lot of people, Bill, that just go around taking the joy out of life. Well, anybody that calls me anything but 'Marine' is going to hear from me. 'Marine' is is good enough for me.

Bill, you never were very literary. But did you ever hear me speak of Kipling and what he said about the female of the species being more deadly than a triple titration of TNT? Well, if a regiment of Marines can make the Germans stand on their bone heads and yell 'Kamerad,' you can imagine what a regiment of female Marines would do? Why, those plop-eyed, yellow-skinned bounders would run so fast and furious they'd never stop for second wind until they reached Berlin.

I never received that German helmet. Are you sure you got the fellow, Bill?

I can't sign myself as affectionately as I used to, Bill. You understand, I'm a soldier now and you wouldn't want me doing anything that wasn't in the Manual.

Yours till the cows come home,

Pvt. Martha L. Wilchinski, M.C.R. (26)

From the very outset Marine Corps policies and expectations were made unquestionably clear to the aspiring applicants. The women enrolled were advised in no uncertain terms that the same rules and regulations governing enlisted male Marines would also govern enlisted women and that failure on the part of any woman to maintain the high standards expected of a Marine would result in the individual being "summarily disenrolled." (27) Colonel Albert S. McLemore made doubly sure that the women to whom he personally administered the oath understood the seriousness of their pledge of allegiance when he warned that he "wanted it distinctly understood that there was to be no flirtatious philandering with the enlisted men at headquarters on the part of the female reservists." (28)

Continually throughout the months between August and November small groups of Marine reservists (F) arrived at Headquarters Marine Corps from cities across the country Although they were dressed in civilian clothes looking like any other young ladies of the time, they were expected to conduct themselves with a degree of military bearing. Where Joint Travel Orders were issued, one woman was placed in charge and was instructed to represent the group and to handle any problems arising during the course of the journey. Meal tickets were supplied by the individual recruiting stations in sufficient quantity to cover the trip, and travel tickets were issued for those who traveled by rail. Sergeant Margaret L. Powers, who was one among five enrolled in Boston on 18 September 1918, remembered how excited they were when they were met at Union Station and whisked off to Headquarters in a mini-bus-like motor car called a "Black-Maria." (29)

Although Headquarters endeavored to have a representative on hand to meet each band of reservists, the effort was not always successful and in those cases the women took the initiative without hesitation and reported in on their own to Headquarters Marine Corps which was located in the Walker Johnson Building at 17th and New York Avenue in downtown Washington.

Name

Marine Reserve (F) was the only official title by which the Corps first enlisted women were known, however, throughout the duration of their service many nicknames were also coined to identify them. On the occasion of their first official visit to Quantico on 21 November 1918, Corporal Elizabeth Shoemaker heard the title "Lady Hell Cats" used for the first time when an enthusiastic Marine shouted it from the crowd as they marched by. During a party planned for the women that same evening, Corporal Shoemaker recalled overhearing one disgruntled young Marine telling his buddies: "This is a fallen outfit when they start enlisting skirts," hence "skirt Marines" was added to the growing list. (30)

The most popular and most widely used of all the nicknames was "Marinette." "The United States Marine Corps frowned upon the use of the word 'Marinette,' " remembered Corporal Avadney Hea, "they posted notices every once in a while on the bulletin board, that we were not to be referred to as 'Marinettes.' We were United States Marine Corps Reserves with 'F' in parenthesis after indicating female. And we were not to be called 'Marinettes.' The Marine Corps didn't like it." (31) In spite of that fact, however, many people still refer, although erroneously so, to the Marine Reserves (F) as Marinettes.

Housing

For those not already living in Washington, the first order of business after reporting in was to arrange for housing. Although the military maintained a limited number of government barracks, they were already filled to overflowing with permanent male personnel and a constant stream of transient troops, so the women were given an allowance for subsistence and quarters (approximately $83.40 per month) and assisted in locating quarters in the city. Private Ingrid Jonassen, who was enrolled in New York on 14 September 1918, remembers how her group enthusiastically bought the Sunday paper their first day in Washington and threw it all away but the classified ads: "We were bound and determined to stay together and continued to look, fruitlessly, all that Sunday. In the end, however, we were forced to hire singly." (32)

The housing shortage by 1918 had reached critical dimensions in Washington and even with the referrals supplied by Marine Corps Housing suitable rentals were hard to find. Complicating matters even worse was the fact that a deadly strain of Spanish influenza was sweeping the city that summer leaving hundreds of dead in its wake. "If anyone even coughed or cleared her throat she was suspect," recalled Sergeant Margaret Powers. (33) The women ended up scattered throughout the city, some found rooms alone in private homes and boarding houses while others in small groups rented apartments.

Cooking presented little problem for those who rented apartments or boarding house rooms, however, meals were rarely included in the single room rentals and most of these women ate in cafes or restaurants in the city. The effects of the influenza epidemic could not be escaped here either, though, and finding available restaurants during the summer months was at times a problem: "We would have to go in and stand third behind somebody who was seated and wait for the two ahead of us to be served, eat and leave before we could be seated," commented Corporal Avadney Hae. (34)

When the housing dilemma was brought to the attention of Mrs. John D. Rockfeller, a member of the War Council, and an enthusiastic advocate of women in the military, additional help was soon on the way. Through her efforts about seventy women Marines were quartered in the Georgetown Preparatory School located in Rockville, Maryland, which although newly built had not yet been opened. Sergeant Margaret Powers who was enrolled in Boston was one among the lucky seventy: "It was a beautiful place with private rooms, and excellent food prepared by a former tearoom owner." (35)

Living in the city had its advantages, however, and the women who were fortunate enough to find quarters near downtown Washington enjoyed a short walk to work while others who were located in the suburbs of Maryland and Virginia rode the trolley cars which faithfully clanged and whistled along their daily routes into the city.

In general the girls found Washington residents to be people of great kindness, such as the lady in Maryland who temporarily set up her sun porch "dormitory style" to accommodate a somewhat discouraged group of six after their long day's search. (36) Sergeant Ingrid Jonassen will also never forget her landlady, Mrs. Pheobus, who helped her through a serious bout with the Spanish flu. (37)

Looking trim in their new uniforms (l to r) PFCs Mary Kelly, May O'Keefe, and Ruth Spike pose at Headquarters Marine Corps. (NatArch RG 127G Photo 530266).

Members of the Adjutant and Inspector's office staff pose for a picture in October 1918; in the foreground are Privates Bess Dickerson (left) and Charlotte Titsworth. (Photo donated by Mrs. Charlotte (Titsworth) Austin).

In spite of the fact that the women were spread far and wide across the city they none-the-less maintained a strong feeling of comradeship with one another and whenever possible were eager to be together as a group, whether at social functions, drilling on the Ellipse behind the White House, or participating in parades and rallies.

Uniforms

As women organized into groups in their effort to support the war, hundreds of unusual uniforms began popping up across the country. Among the most admired, by far, was the uniform designed by the Quartermaster for the Marine Corps Reservists (F). "The civilian girls were terribly envious," recalled Ingrid Jonassen, "and tried on occasion to borrow my uniform." (38) The girls looked stylish indeed in their tailormade two-piece suits of green wool in the winter and tan khaki in the summer. Accessories included: a specially designed shirt, regulation necktie and overcoat, brown high-topped shoes for winter and oxfords for spring and summer - the first low shoes most of them had ever worn. Matching overseas caps on which they proudly wore the globe and anchor insignia were the preferred headgear; however, during inclement weather the women reservists were permitted to wear campaign hats which made "excellent umbrellas," recollected Private Florence Gertler "and were really quite comfortable and practical when it rained." (39) Raincoats, gloves, and purses were not issued, and while old photographs show some women wearing gloves, handbags were not permitted on any occasion. (40)

Although initially each woman had only one winter uniform set, the standard uniform issue was established as: two winter and three summer uniforms, which, in addition to jackets and skirts, consisted of six shirts, one overcoat, two neckties, two pair of high topped shoes, and two pair of oxfords. The women did not have a dress uniform. As individual items became unserviceable through "fair wear and tear," the Quartermaster's Department replaced them on an exchange basis (41)

The uniform regulations governing women reservists were no less strict and no less important than those governing male Marines and most of the women took great pride in keeping their uniforms "spick-and-span" and their shoes polished to Marine Corps standards. "Even though the uniform regulations were strict," commented Private Mary Sharkey, "the women used their ingenuity and were put to relatively little inconvenience":

Private Carrie E. Kenny pictured in her World War I uniform in 1919; she continued to serve the Marine Corps as a civil servant for 26 years following her release from active duty. (NatArch RG 127G Photo 13458).

The coat or blouse had a large pocket on each side which extended across the front. These commodious cavities were handy, but putting too many things in them created bulges which was strictly against regulations. To avoid this I kept a comb and brush in my desk, tucked my handkerchief inside a sleeve, and carried a small change purse so as to appear trim and neat. There was no need for a wallet. Our underwear and stockings we bought ourselves. The skirts were heavy enough not to require a slip, but if one wore a slip it tended to bunch up in front while walking or marching, so I wore long bloomers over my underpants. They stayed in place nicely. The skirt had to reach the tops of our high-tied, pointed-toed, curved-heeled shoes. Our overseas cap was made with a fold that could be released and buttoned under the chin, thereby covering hair and ears. The cap was of the same material as the uniform. My quarters had laundering and ironing facilities, so I encountered no difficulty in maintaining cleanliness...Initially we each had only one winter field uniform and when it was sent out to be dry cleaned we requested permission to wear civilian clothes for that day only. (42)

Although the Navy women also had uniform regulations, they were quite a bit less strict than those governing Marine women reservists "The Navy Department girls were quite a bit more feminized," remarked Corporal Elizabeth Shoemaker. "Enlisted women in the Department were not subject to the same discipline as were the Marine girls. For instance, while the Navy girls wore lace collars and jewelry, the Marine girls, even on the hottest Washington summer day, were not allowed to turn down their collars or remove their fore-in-hand neckties." (43)

Violations of the uniform code while only occasionally disciplined were continually brought to the attention of the women reservists. The following memorandum written in 1919, along with a rather unusual violation in the wearing of a ribbon of high distinction, points up the more common infractions for which the women were reprimanded:

HEADQUARTERS U.S. MARINE CORPS
Adjutant and Inspector's Department
Washington, D. C.

February 8, 1919

MEMORANDUM FOR: THE ADJUTANT AND INSPECTOR,
THE PAYMASTER,
THE QUARTERMASTER

It has come to the attention of the undersigned that there is a total disregard of uniform regulations on the part of a number of female reservists attached to these Headquarters. Several of them have recently been seen in public places wearing rifle insignia, Sam Browne belt, spats and non-regulation shoes. A flagrant example of this disregard of regulations was noted today in the case of a female reservist who was wearing a ribbon of the Croix de Guerre on her coat in the position required by regulations for the wearing of campaign ribbons and ribbons and medals. This is not only a direct violation of uniform regulations but is as well the wearing of a decoration which is awarded by the French Government only for gallantry on the field of action in the face of the enemy.

Female reservists must be made to understand that the wearing of the uniform by them confers no priviledge to deviate from the uniform as prescribed in regulations.

By order of the Major General Commandant. (44)

The most popular uniform item among the women reservists was the Sam Browne belt. Although the girls were warned repeatedly against its wearing, "It made the uniform look particular stunning," reminisced Sergeant Jonassen, so most of the women willingly took the risk and wore it whenever they could, on leave and liberty, at rallies and parades. On one occasion a young woman was seen wearing the belt by the major for whom she worked and the next morning she found herself standing tall being sternly reprimanded with a warning forbidding her to wear it again. "Some months later, recalled Corporal Elizabeth Shoemaker, "a great victory parade was held in Washington. General Pershing led it, the girls were in it, and of course all the Marine Corps officers...The day before the parade the orders for dress were posted directing all officers to wear Sam Browne belts. A mad scramble in the Washington stores to buy belts followed soon after and the stores were quickly sold out. The major who had been so cross at his secretary could not find a single belt in the whole city...An hour before the parade he rang for her and asked if he might borrow hers, he even sent her home in an official car to get it." (45)

Private, later Corporal, Avadney Hea dressed in the favorite uniform of the Marine Reservists (F) - winter field service green, with Sam Browne belt and swagger stick. (Photo donated by Mrs. Paul (Hea) Cheney).

The women were particularly fond of having their pictures taken in the winter field green uniform, wearing the Sam Browne belt and holding a swagger stick: "We were permitted to carry a swagger stick," remembered Corporal Mabelle Musser, "which some of us did with high glee - not every day to work, but on special occasions and on Sunday when we went strutting." (46) Eventually, although officially frowned upon, a liberal policy was adopted with regard to the Sam Browne belt and the reservists when not on duty were permitted to wear it with relatively little repercussion from headquarters.

The women assigned to Washington, D.C. were the first to be completely outfitted in their uniforms since they were able to take immediate advantage of local tailors and the Quartermaster's supplies. Women recruited for duty in other cities, however, waited months and in some cases did not receive uniforms at all. Private Lucy Ervin, who was assigned to the District Division in Indianapolis, Indiana, after having waited several weeks could wait no longer and ingeniously assembled her own uniform:

> I could see that it would be months yet before I could be assured a uniform and since I had always admired the dress uniform of the Marines, I was given permission to wear one of the dress jackets, to which I added a black velvet skirt, web belt complete with red and gold stripes and a campaign hat. I'll tell you I was high off the ground when I walked down the street in it...Many officers were surprised to be given a salute as I passed them...I was just a private in the rear ranks but I was proud to be doing a job that was needed." (47)

In a letter to Headquarters Marine Corps following her discharge Private Ervin told her delightful story and by return mail she received a box from the Quartermaster with material and emblems for a regulation winter field uniform.

After the war when asked about the uniforms worn by the Marine Reservists (F), the Secretary of the Navy, Josephus Daniels, commented:

> The uniforms...of the Marines (F) were natty and beautiful, were worn with pride...and suited to the duty assigned. As a designer of women's uniforms the Navy Department scored a distinct success, for these uniforms were copied by women all over the country." (48)

Pay

The 305 women reservists enrolled between the months of August and November 1918, proved to be a capable and industrious group and by the end of the war a good number had been promoted to private first class, corporal, and sergeant, the highest rank obtainable. In keeping with Marine Corps policy, the women received the same pay as enlisted men of corresponding rank: private, $15.00; private first class, $18.00; corporal $21.00; and sergeant, $30.00 per month, plus an additional $83.40 per month for subsistance and quarters. (49) The end of the first payroll period held quite a nice surprise for Private Ingrid Jonassen: "I never expected to be paid for my services," reminisced Sergeant Jonassen, "I thought that we would receive only room and board, so you can imagine my surprise when my first pay check came through." (50)

The women had to work hard to prove themselves and promotions, given for time in grade, time in service, "work well done, and conduct becoming a woman," were worn with the greatest pride and pleasure. (51) Private First Class Ida Kirkham recalled the unexpected problems she incurred with her initial promotion:

> After a few months I received my first promotion to private first class. My stripe, which reminded me of a whisk broom, earned me $3.00 more a month and entitled me to be called "Private." I remember how devastated I was when I was informed that my whisk broom was sewn on upside down. After all, no one had instructed me how to wear that stripe and I sewed it on the way I thought it looked best." (52)

By the war's end the senior enlisted woman with the rank of sergeant was Opha Mae Johnson, who worked in the office of the Quartermaster,

Assignments To Duty

The jaunty women reservists arrived at Headquarters Marine Corps eager for work and little time was wasted preparing them to take the jobs of Marines anxiously awaiting overseas orders. One by one they took their places as secretaries, office clerks, and messengers in the offices of the Adjutant and Inspector, Paymaster, and Quartermaster. The muster roll for Headquarters Marine Corps dated January 1919, show the distribution of women reservists to the various departments as follows: (53)

Members of Parris Island's Drum and Bugle Corps show Corporal Martha Wilchinski how to execute a drum roll during a visit to the base in 1918. (NatArch RG 127G Photo 518919).

Women Marines post recruiting posters in New York, (l to r) Privates Minette Gaby, May English, Lillian Patterson, and Theresa Lake. (NatArch RG 127G Photo 530751).

	Sgts	Cpls	Pvt Fst Cls	Pvts
Adjutant and Inspector .... (BGen Charles H Lauchheimer)	1	16	17	54
Quartermaster .... (BGen Charles L. McCawley)	1		8	44
Paymaster .... (BGen George Richards)		3	6	19

In most cases the superior qualifications possessed by the women reservists insured them excellent job assignments. In the office of the Quartermaster, Corporal Mabelle Musser was detailed as a secretary to the department handling motor transport assignments, Private Marie Schlight was put to good use as secretary to the officer in charge of Marine aviation, and Sergeant Margaret Powers was so skillful that she was immediately assigned as secretary to Colonel William Fay in the office of the commandant.

Corporal Louise Hedtler, secretary to Captain Charles Ketchan in the Adjutant and Inspector's office, researched items in connection with congressional inquiries: "It was the responsibility of the Adjutant and Inspector to give congressmen every bit of information Headquarters Marine Corps could, even going so far as writing units overseas in response to their inquiries," she remembered. (54) As an additional job Louise and Private First Class Florence Gertler, secretary to Captain Francis C. Cushing, Assistant Adjutant and Inspector, were given the responsibility of sending Marine Corps casualties' personal effects to next of kin with an appropriate letter of condolence: "Each case was unique" remembered Louise Hedtler, "and Major General Commandant Barnett read and signed each letter separately before it was forwarded to the family." (55) So professional was the work of these two women that General Barnett summoned them in order that he might personally compliment them on their "performance of duty." (56)

Sergeant Martha Wilchinski, Private First Class Mary Benson, Private First Class Mary Kelly, Private First Class Mary Reilly, and Sergeant Lela Leibrand were assigned to the Marine Corps Publicity Bureau, where they performed general office work and, in the case of Sergeants Wilchinski and Leibrand, contributed articles for publication in the Recruiters' Bulletin, Leatherneck newspaper, and The Marines Magazine. Sergeant Wilchinski, who came to the Marine Corps well prepared with a degree in journalism from New York University went on to become the editor of Variety Magazine and Sergeant Leibrand, who made the first military training film, "All in a Day's Work," pursued a career in filmmaking at Fox Studios in New York following her discharge. (57)

A small number of women who looked "particularly sharp" in their uniforms were assigned as messengers to make trips throughout Headquarters Marine Corps and various offices of the Navy Department with official communiques and intra-office correspondence. These girls, among whom were Privates First Class Mary O'Keefe, Ruth Spike, and Samia Ellyson, were reputed to be the most representative among the Marine Reservists (F).

In addition to their assignments in Washington, special permission to enroll women was also granted to recruiting officers who were severely pressed for additional help in the cities of New York, Rochester, Boston, Philadelphia, Indianapolis, Denver, Spokane, San Francisco, and Portland, Oregon. The women were to be recruited locally and put immediately on the job to free as many permanent male office personnel for reassignment to the front as possible and to help out with the mountain of paper work inherent in enlisting large numbers for military duty. Private Elizabeth Bertram, who was working as a legal secretary, was enrolled for duty in her hometown of Denver, Colorado. So enthusiastic was she about being a Marine that when the month arrived during which the Denver Marines were required to qualify on the rifle range, she was determined not to be left out, which posed an unusual problem for Major Charles Guggenheim, the officer in charge of the Denver Recruiting Station:

> I grew up in a small town and in my younger days had done some hunting and a great deal of target practice. I considered myself a better than average shot so, when I learned that qualifying on the rifle range added dollars to pay checks, I asked for permission for a tryout. The major gave me an unequivocal "no," the official firing range was on the western slope of the Rocky Mountains more than a hundred miles away and he certainly would not send a female on such a jaunt simply on her say so that she knew how to handle a gun...Major Guggenheim turned to Corporal Jones who worked in the office, 'She thinks she can shoot' he said. 'Day after tomorrow you will take her to the Denver Police practice range just outside the city limits. Make all necessary arrangements for a one person regulation full course tryout.'
>
> 'Day after tomorrow' was a cold, windy, overcast day with squalls of icy rain; but orders were orders so Corporal Jones and I, equipped with blank target sheets, ammunition, and an Enfield Army rifle went to the designated place. The rifle range was a vast expanse of clayish mud with a few outcroppings of rock. There were several policemen waiting under the shelter to see what was going on. The rifle was

too long in the stock to hold a firm pressure against my shoulder, there were surpressed snickers from the kibitzers. With almost 20/20 vision I did all right on the standing positions and thought I even made a bullseye. I did fairly well on kneeling and sitting, and managed a few good shots prone; but every time I tried rapid fire the result was disaster. I stuck it out for the full course but my shoulder was pounded almost into jelly and my right arm was a painful something that didn't seem to belong to me. We returned to the office sopping wet. My hair was stringing down from under my dripping cap, my clothes were plastered with mud and even my face was dirty. The major sent me home in a closed car without comment. The next day again spic-and-span but with the grandfather of all bruises spreading over my shoulder, arm and even on my chest, I was summoned to The Presence. My target sheets were spread over the desk with every bullet hole carefully evaluated. 'You seem to have qualified unofficially' the major acknowledged. 'You will be included in the next class to the official range.' Then he arose from the desk and approached me saying 'In the meantime you may wear this,' and he pinned his own engraved marksman medal on my blouse. I intended to execute a smart salute in acknowledgement; but when I tried to raise my right arm I involuntairly yelled 'Ouch!' instead, and ran to my desk in an agony of fear and shame...Major Guggenheim, the strict disciplinarian, he of the hair-trigger temper and the blistering vocabularly shot one glance at the astonished and apprehensive faces around him, and saved the day with a burst of hearty laughter. (58)

Private Bertram, unfortunately, was never able to go to the official firing range. On 11 November 1918, the Armistice was signed and she along with the rest of the Marine Reservists (F) was transferred to inactive duty.

In addition to doing the clerical and routine office work at the recruiting stations the women were also used as recruiter's aides. They were reported to be very impressive with their "trim and snappy appearance and their business-like attention to their work." (59) The women were effectively used for publicity in bond drives and rallies, photographs, and as members of recruiting teams touring the country with Marine Corps exhibits. Sergeant Martha Wilchinski was used frequently in publicity shots. Although women were not assigned on board ship or as members of the Marine Band, some of the more widely circulated photographs

Standing rigidly at attention behind their drill instructors women Marines prepare for a morning's drill on the Ellipse behind the White House. (WarDept Photo 165-WW-598A-11, NatArch).

Following a morning's drill session, these Marines took time to pose for a platoon photograph before continuing on to their offices to begin the day's work. (NatArch RG 127G Photo 530165).

of Sergeant Wilchinski show her going through the daily routine with sea-duty Marines and members of the Marine Corps Drum Corps.

Corporal Mary Kathryn Beck, Rochester's only female Marine, was assigned as a member of the recruiting team in that area. During a combined Navy/Marine Corps exhibit in Rochester, the Recruiters' Bulletin reported that "Corporal Beck...dressed in her snappy uniform - Sam browne belt, overseas hat and swagger stick...was no small part of the success of the show." (60)

In most cases the male Marines were very protective of the women and helped them as much as they could in the performance of their jobs. Although some of the women came up against negative attitudes harbored by Marines who felt that the Marine Corps "stepped down" when it enrolled women, most of the female reservists, regardless of where they were stationed, felt that they were treated as equals: "The men did not look down or frown upon us," recalled Private First Class Edith Macias, "actually they were glad to have us. We were given a job to do and we did it. We were definitely not considered decorative rather than practical, but were treated as professionals." (61)

Drill And Parades

A very important part of the day for the Marines Reservists(F) was spend at 7:00 o'clock in the morning on the Ellipse behind the White House. With stern sergeants barking commands of "forward, march," "to the rear," and "platoon, halt," they drilled in earnest. "Initially," remembered Private First Class Edith Macias, "the male drill instructors were indignant to have been selected to teach drill to women. As a result they showed us no mercy and taught us the same way as they did male recruits." (62) Private First Class Ida Kirkham recalled how her drill instructor went along the line ordering them into the correct positions with commands such as "stick out your chest," and "suck in that gut." (63) "They were perfectionists, even stinkers," stated Private First Class Florence Gertler, "but they were good." (64)

Most of the women found drill "fascinating" and they tackled the challenge with energy and enthusiasm. Some did better than others, however, and while there were those who responded almost immediately to the unfamiliar sound of the drill commands there were still others who never did quite get the knack of it. "We all loved it," reminisced Corporal Elizabeth Shoemaker, "but we couldn't do it...We had several drill instructors and they all tore their hair and went crazy. Some even pleaded with us and tried to bribe us to do better." (65)

Marine Reservists (F) followed by Navy Yoemen (F) marching in one of the many parades in which they took part in Washington during 1918-1919. (NatArch RG 127G Photo G-521222).

Marine Reservists (F) march smartly down Pennsylvania Avenue in Washington when the President reviewed the 2d Division on its return from France in 1919. (NatArch RG 127G Photo 528255).

The women practiced hard under the critical eye of their instructors and were soon regular participants in the numerous parades held during the months just before and after the signing of the Armistice. When the 2d Division came home from France in 1919, the women were positioned behind the wounded Marines and "Marched as proud as peacocks" down Pennsylvania Avenue in an unforgetable "Welcome Home" salute to America's fighting men. In addition to marching as a unit in the parade the women were also on duty at the reviewing stand: "The parade was reviewed by President Wilson," recalled Corporal Sarah Jones. "The reviewing stand was at Lafayette Park, across Pennsylvania Avenue from the White House, and the Marine Reservists (Female) formed part of the honor guard standing in front....just below President Wilson." (66)

In most parades and ceremonies the women were placed in positions of honor, either behind the Marine Band or as members of the honor guard. When the body of the Unknown Soldier was returned from France, Sergeant Ingrid Jonassen was chosen to be a member of the solemn procession: "I walked directly behind the caisson on the way to Arlington," she recollected. It is interesting to note that Sergeant Jonassen reported that she often has seen herself in old newsreel clips still shown on television of that ceremony.

On 3 February 1919, the women received a special honor when Major General George Barnett personally reviewed them on parade at the Ellipse. Eugene J. Cour, noted director of the International Film Corporation, was on hand to take motion pictures of that unique moment in military annals. Their "snappy appearance" so inspired Mr. Cour that he subsequently entitled his feature film story, "No wonder the Kaiser Quit." Private Avadney Hea remembered how the girls flocked down to the moving picture theatre the following week "to see ourselves in uniform." (67)

Due to the wide dispersion of the women throughout the city in accommodations ranging from apartments to private homes there was no formal quarter's inspection. They did, however, stand in-line inspection given initially by the noncommissioned officer in charge, and later by the senior woman, which took in "posture, hair style, uniform, shoe shine, and all the rest, with demerits given for infraction of the code," reflected Private Mary Sharkey. (68)

PFC Edith Macias, one of two sisters who served as Marines in World War I, during a review of Marine Reservists (F) by the Major General Commandant. (War Dept Photo 165-WW-598A-10, NatArch).

Discipline

Discipline for the women reservists while not as rigid as that governing male Marines was still very much Marine discipline. The women were expected at all times, on duty as well as in their free time, to conduct themselves with the dignity and military bearing befitting a member of the Corps. It did not take the Reservists (F) long to realize that if they were to be equated the same privileges as their male counterparts they would also be expected to take their share of the duties. An incident remembered by Corporal Elizabeth Shoemaker aptly points up this fact:

> After we had been at Marine Corps Headquarters for two weeks a typical hard-boiled sergeant who just loved to do it, ordered us to sweep the floors and wash the windows in our offices. Two pretty girls, from wealthy families, rushed to Colonel McLemors's office and said that they would not undertake such labor. He was very angry with them, reminded them that they had enlisted, that they couldn't change their minds about their duties. He ordered them to wash those windows and they did it. It was wonderful discipline for young girls and we came to love the Marine Corps better than anything in life. (69)

Their conduct was under constant and close scrutiny and in the following letter dated 8 March 1919, Major General Barnett reminded his department heads of their responsibility to ensure that the traditionally high standards demanded by the Corps were being upheld by enlisted personnel under their command:

"HEADQUARTERS U. S. MARINE CORPS,

Washington

March 8, 1919

From: The Major General Commandant
To: The Adjutant and Inspector,
The Paymaster,
The Quartermaster.

Subject: Conduct of enlisted personnel

1. Enlisted men and women on duty at these Headquarters have been observed skylarking

together on the streets, and not infrequently have been noticed walking arm in arm.

2. Please bring this matter to the attention of the enlisted personnel in your departments, informing them that such practices must cease and that they must at all times conduct themselves in a dignified and soldierly manner.

GEORGE BARNETT" (70)

Rank was a preciously earned privilege to the girls and when it was taken away from them in the course of disciplinary action it was indeed a bitter pill to take. Private Elizabeth Shoemaker whose first furlough home began with all the excited anticipation of a soldier's first homecoming was destined to end on a sour note:

> We had our first furloughs right after we received our uniforms and everybody wanted to go home to show them off...So home to New Jersey I went. They did try at Headquarters to teach us the fundamentals of discipline and Marine Corps decorum, but I guess it left me untouched. I was having such a good time at home, it was just a short furlough, and my mother had planned a party for the day after my furlough was over. So, I sent a telegram to Colonel McLemore saying that I was delayed and wouldn't be back for three days after the end of the furlough. And worst of all...I sent him my regards. When I got back a day or so later there was a message for me to report to Colonel McLemore's office...He informed me that I had been AWOL and gave me what is called a 'deck court-martial'...I was quite shaken until I came out of his office. Through the grapevine word had gotten around and there was a crowd of enlisted men waiting to congratulate me; the Marine Corps newspaper published the shameful story and I got letters from Marines all over the world, including China, telling me not to worry, and saying "Now you are a real Marine." The girls could not hold commissions but could be sergeants. Because of my court-martial I never became a sergeant, but remained a corporal. (71)

Occasionally the women were disciplined by their hard driving drill instructors for ineptness at drill. Private First Class Florence Gertler, who was the pivot of her platoon, related how she caught her heel in the soggy dirt one rainy morning and was unable to turn, "I got KP for that! I sure must have looked funny with my body going in one direction and my leg remaining in the other." (72)

KP for the women usually consisted of a Saturday afternoon washing office windows, sweeping floors, or cleaning bathrooms at Headquarters Marine Corps.

Social

Although the girls worked hard, some on a 10-hour day, six-day week schedule, they found time to relax and enjoy their many new friends, and the interesting sights and sounds of the Nation's capitol. "Usually the girls who lived together had their social life together," recalled Private First Class Florence Gertler. "The big event of the week was Sunday night supper. Our room and board did not cover the Sunday evening meal and since we were usually broke, particularly toward the end of the month, if one of us was asked out for a Sunday supper date we always tried to ring in a friend by double-dating." (73)

According to most of the women none of the girls wanted for dates, "any fellow was proud to take out a female Marine Reserve," reminisced Corporal Louise Hedtler. The girls were true to the Corps, however, and the Marines always took first place. Sergeant Ingrid Jonassen recalled that "every girl was begged for at least two dates every night." Overzealousness caused Sergeant Johnssen on one occasion to make seven dates for the same evening. In a panic, unable to recruit other girls, she asked her landlady, Mrs. Canicius if she could invite them to the house. They all came and enjoyed Ingrid's company and an evening of Mrs. Canicius' home cooking. (74)

The overnight expansion to gear up for war caused Washington to take on a boom-town atmosphere and the city literally hummed with activity. "Washington was a dancing town," recalled Corporal Elizabeth Shoemaker. "There were a lot of roof gardens on top of the hotels and we danced our heads off...to tunes like 'The Japanese Sand Man' and 'Jada'." (75)

The first official visit of the Marine women to Quantico on 21 November 1918, was occasioned by the invitation extended by the staff of the Leatherneck newspaper. As the 13-woman delegation stepped off the train in Quantico town to the strains of the post band playing "Hail, Hail the Gangs All Here," and "How Do You Tame Wild Women," the crowds of waiting Marines cheered with enthusiatic approval. "The camp to a man was knocked cold," reported the Leatherneck:

That's all there was to it. It was just a plain case of 'Knock'em for a row of Hostess houses.' They lined up in a column of twos and marched behind the band to the parade grounds where they watched the O.T.C. (Officer Training Class) men parade. They saluted colors and retreat...Them they made the offices of The Leatherneck their headquarters. Boys, they are just like any other girls in some respects, for they carry mirrors with Marine emblems on them, but they are regular Marines, for they are very proud of their personal appearance and carry their own powder.

And how those Marine girls can dance!

Everyone was proud of the Marine girls. They carried themselves like real soldiers, were good follows and proved that they were ready to go anywhere and conduct themselves with honor to the Marine Corps. (76)

The lucky thirteen who termed this occasion "one never to be forgotten, and entrapped in our memories forever" were: Mae E. Bennett, Samia Ellyson, Mary Clay, Mary L. O'Keefe, Katherine Pidgeon, M. Helen Fell, Charlotte M. Shoemaker, Ina C. Tracy, Martrese Thek, Mary C. Kelly, Edith Macias, Sarah Macias, and Jeannette Roumage. (77)

This first trip to Quantico was just one of many to follow for the Marine girls. Dances were scheduled every Friday night and those who lived in the city commuted easily by train from Union Station. After the Armistice, Private First Class Florence Gertler and her group of friends used to attend the dances whenever they could. "Usually the trains were so crowded we'd wind up sitting on the floor or lying down on the train platform to get some sleep." she reminisced, "There were civilian girls as well as other military women who came, but every Marine girl was assured of all dances, even though some of the other girls might not." (78)

The girls were competitive in nature and, typical of Marines throughout the ages, were eager for a little friendly jousting between departments. "Marine Girls to Have Own Team," read The Leatherneck headlines:

The girls of the Marine Corps at Washington have organized a basketball team to play rival teams of the Army and Navy Departments. All of which sounds sort of formal and everything to you men here at Quantico

Marines on board a commemorative float before taking their place in the parade line in Philadelphia in 1919. (Marine Corps Museum Photo Collection).

Secretary of the Navy Josephus Daniels and Assistant Secretary Franklin D. Roosevelt pose with Marine Reservists (F) and Navy Yoemen (F) at their final review in 1919. (NatArch RG 127G Photo 530164A).

> but listen to the lineup announced by "Acting Jack" Mae E. Bennett. There's "Caddy Kid" Okey, "Cutey" Montgomery, "Gin" Bellinger, "Sammie" Ellyson, "Fido" Shoemaker, "Stonewall" Bennet, "Martrese" Thek, "Shorty" Kelly, "Tubby" Tracy, "Jane" Macias, "Red" Barker and "Priessy" And they have the sweetest little uniforms too. (79)

According to Charlotte Titsworth, who also played on the team, they wore "navy blue turtle neck sweaters with USMC in white across the chest." (80)

The girls also showed their prowess as first rate entertainers when they planned, organized, and executed a vaudiville show for the troops at Quantico. The <u>Leatherneck</u>, a great supporter of the girls in green, ran a feature story advertising the extravaganza. "A REAL LIVE SHOW INCLUDING BURLESQUE ON APACHE DANCE AND ENTICING CHORUS OF MARINE BEAUTIES NOW REHEARSING," broadcast the paper.

The staring number, "Shades of Cleopatra's Jazz" was planned to be a real show stopper: "The girls will...spring it on their unsuspecting Buddies January 20 with a full line of special music, a beauty chorus of dancingly dainty yet doughty and alluring Marine girls soubrettes who will attire themselves in semi-Cleopatra costumes. All they will lack will be the river Nile." stated the article.

The show was such a success that the following year, seven months after the women had been deactivated, Washington's Belleau Wood Chapter of former Marine Reserves (F) returned for another appearance, this time in two one-act comedies, "'Op' O' My Thumb" and "The Suburbanite." (81)

Disenrollment

With the signing of the Armistice of 11 November 1918, the urgent need for the support of female labor gradually decreased and finally on 15 July 1919, orders were issued by the Major General Commandant for the transfer of "all reservists on clerical duty at Headquarters...to inactive status prior to 11 August 1919." (82)

With traditional Marine Corps pomp and circumstance, a great ceremony was arranged on the occasion of their departure from the Corps back into civilian life. Major General Barnett accompanied by his complete staff and the Marine Corps band reviewed the women for the last time on the White House lawn. As the women completed their final pass-in-review Josephus Daniels stepped up to give the farewell address. Corporal Elizabeth Shoemaker vividly recalled the unforgettable occasion:

> We stood in front of him in our uniforms listening to every word of his eloquent speech; He said we had been good Marines and he was proud of us; Then, in his closing statement he said, "We will not forget you. As we embrace you in uniform today, we will embrace you with out uniform tomorrow." All down the file of men standing at strict attention, the line broke, and everyone roared with laughter. The Secretary of the Navy forgot he was talking to women. (83)

Disenrollment from the Marine Corps Reserve (inactive status) took place at a steady, but gradual, rate and evenually all women were discharged although this was as late as 1922 in some cases. (84)

On 30 June 1919, Major General Commandant George Barnett was asked to comment on the service rendered by the female reservists during the war. In reply, the Commandant stated:

> It is a pleasure, but not by any means an unexpected one, to be able to state that the service rendered by the reservists (female) has been uniformly excellent. It has, in fact, been exactly what the intelligence and goodness of our countrywomen would lead one to expect. (85)

Benefits

As veterans of the war the Marine Reservists (F) were afforded the full benefits legislated into law, the same as their male counterparts. In addition to the coveted right to military burial in Arlington National Cemetery alongside others who have fought to preserve America's honor, the women were also eligible for government insurance, World War compensation, a $60.00 bonus upon discharge, medical treatment and hospitalization for service-connected disability, and a five per cent bonus added to Civil Service examinations for those who chose to remain in government service. (86)

With typical Marine esprit, some of the reservists elected to remain on duty as civilians at Headquarters Marine Corps, supply depots, and recruiting offices across the country. Still others returned to their hometowns and the prewar positions they so courageously left when the appeal for volunteers was made.

Lieutenant Colonel Martrese Thek Ferguson, who served as a woman Marine in both World Wars I and II, is pictured in November 1945. (USMC Photo A412893).

At the conclusion of the war, the women who remained in the reserves in an inactive status received retainer pay of $1.00 per month until the end of their four-year enlistment contract. In addition they received a Good Conduct Medal and World War I Victory Medal when they were discharged from the inactive reserves.

After The War

Among those who remained in government service was Mrs. Jane Van Edsinga Blakeney who became head of the Decorations and Medals Branch at Headquarters Marine Corps, serving three commandants in that capacity. In 1959, Mrs. Blakeney was honored by the Marine Corps Combat Correspondents Association for her monumental book, Heroes, U. S. Marine Corps, 1861-1955, which lists the medals earned by Marines during that 94-year period and records the names and citations of those medal winners.

Another woman who chose to remain was Private Alma Swope who was to serve in the Supply Department for more than 44 years. Private Swope has the distinction of being the last Marine Reservist (F), who accepted a civil service position, to retire from government service. Upon her retirement in 1963, she received the personal congratulations of General David M. Shoup, then Commandant of the Marine Corps, and a letter of appreciation from the Secretary of the Navy.

In the years following the war, Corporal Sarah Macias, Private First Class Edith Macias, and Private Mary Sharkey, contributed to the Marine Corps in a very special way. Sarah's only daughter, Major Anne Lenox, serves today as a member of the Marine Corps Reserve, Edith was to have four sons and one step son serve the Corps, and Mary proudly claims three sons who wore the scarlet and gold.

Corporal Leola Hopkins can not only boast that her daughter, Leola, was the first woman to enlist in Norfolk during World War II, but also that she is still represented in the Corps by her other daughter Nannette, who is a reserve officer.

In an effort to maintain contact with the military and each other, many Marine Reservists (F) joined associations such as the Marine Corps League, Women Marines Association, and the American Legion.

The slogan "Once a Marine, Always a Marine" was true for two World War I Women Reservists who came back to the Corps to serve as officers in the Marine Corps Women's Reserve during World War II. Martrese Thek Ferguson, who

In 1946, Mrs. Opha Johnson, the Marine Corps' first woman Marine, and former Director of Women Marines, Colonel Katherine A. Towle, admire the uniform worn by Mrs. Johnson, modeled by PFC Muriel Albert. (USMC Photo 313950).

graduated at the top of thé first officers' class in the Marine Corps Women's Reserve, rose to the rank of lieutenant colonel, and commanded more than 2,000 women at Henderson Hall in Arlington, Virginia. The other woman reserve officer who wore the Marine green uniform in both wars was Lillian O'Malley Daly, one of the eight women who came into the new Women's Reserve directly from civilian life in early 1943. She was immediately sent to Camp Pendleton, California, where she served as West Coast Liaison Officer for the new women's organization.

One might be courious about how the women themselves felt about serving the Corps. The sentiments expressed by Private Mary Sharkey describes most accurately the feelings of the other World War I women: "I loved being a Marine. I loved helping our country. The feeling that the Marine Corps and I helped each other will always remain with me... however, the scales were definitely tipped on my side. It was, all in all, a wonderfully gratifying experience." (87)

The Nation's effort to prosecute the "War to end all Wars" marked the beginning of a new era in the history of American women. World War I afforded women across the Nation the opportunity to try their hand at hundreds of new and challenging occupations, many of which were considered unthinkably inappropriate for women before the war. They worked diligently and ably and in so doing laid the foundation for higher wages, better jobs, improved working conditions, and a more competitive status in the American labor market.

Among the more profound issues brought to a head as a result of the participation of American women in World War I was that of the political question of women's right to vote. Their involvement had been on such a large scale and so important to the war effort that the politicians were forced to consider the demands of the women's movement more seriously than they had before the war when the gainful employment of women outside the home was minimal and restricted.

By 1919 the movement's spokesmen had developed a most influential champion in President Woodrow Wilson who repeatedly spoke out in favor of the cause. In a speech to Congress concerning women's suffrage, President Wilson brought the moral implications of the issue into focus when he said: "Unless we enfranchise women we shall have fought to safeguard a democracy which, to that extent, we have never bothered to create." (88)

Former Marine Reservists (F) return to Headquarters Marine Corps for the 51st Anniversary of their enlistment. Standing l to r: Colonel Jeanette I. Sustad, former Director of Women Marines; Miss Elsie Pinney; Mrs. Alma Swope; Sue Seusa, President, Metropolitan Washington Chapter, Women Marines Association; Mrs. Philena Beale; Mrs. Marie Kane; and Sergeant Major Mabel A. R. Otten, former Sergeant Major of Women Marines. Seated l to r: Mrs. Louise McLuckie; Mrs. Marie Buttell; Miss Carrie Linkins; and Mrs. Margaret Jeschke. (USMC Photo A419732).

At length Congress adopted the Nineteenth Amendment which specified that the right of citizens to vote should not be denied by the United States on the basis of sex, and the women of America stepped up to the polls to vote for the first time in the 1920 Presidental elections.

Appendix A

Notes

(1) Lucy Brewer, The Female Marine, Philadelphia, 1917.

(2) U.S. Marine Corps Muster Rolls for 1-31 August 1918.

(3) Ida Clyde Clarke, American Women and the War, (New York: D. Appleton Company, 1918), p. 6.

(4) The New Republic, 20 October 1917, p. 6.

(5) World's Work, 5 June 1917, p. 45.

(6) U.S. Women's Bureau, The New Position of Women in American Industry, Bulletin #12 (Washington, D. C., 1920), p.25.

(7) U. S. Department of the Navy, Bureau of Naval Personnel, Women's Reserve, "WAVES History," Unpublished manuscript, (Washington, D. C., 1946), p.1.

(8) Eunice Dessez, The First Enlisted Women, 1917-1918 (New York: Dorrance & Co, 1946), p.12.

(9) National Archives Record Group 80 -- Records of the Secretary of the Navy, General Correspondence, 28555-510-- Letter dated 2 August 1918. Copies on file Reference Section, History and Museums Division, Subject file: WW I Women Marines, hereafter Women Marines World War I File.

(10) Ibid.

(11) Ibid., 8 August 1918.

(12) Typed excerpt Chapter XXXI, "Women in the Navy," from Josephus Daniels book, Our Navy at War (Women Marines World War I File).

(13) National Archives Record Group 127--Records of the U.S. Marine Corps, General Correspondence (1913-1938) 1535-1555, Circular, dated 14 August 1918 (Women Marines World War I File).

(14) "Women in the Navy," Chapter XXXI, op. cit.

(15) Ibid.; MC Circular dated 14 August 1918, p. 3. (Women Marines World War I File).

(16) *Ibid.*

(17) Interview of Mrs. Florence (Gertler) Miller, p.2, hereafter *Miller interview* (*Women Marines World War I File*).

(18) Interview of Mrs. Ida (Kirkham) Lawrence, p.1, hereafter *Lawrence interview* (*Women Marines World War I File*).

(19) *Miller interview*, p.1.

(20) *Ibid.*, p.3

(21) National Archives Record Group 127, *op. cit.* Letter from MajGen George Barnett, dated 30 June 1919.

(22) Interview of Mrs. Elizabeth (Shoemaker) Linscott, p.8, hereafter *Linscott interview* (*Women Marines World War I File*)

(23) *New York Evening World*, 23 January 1919.

(24) Interview of Mrs. Mary (Sharkey) Short, p.3, hereafter *Short interview* (*Women Marines World War I File*).

(25) *Linscott interview*, p. 6.

(26) Kemper F. Cowing, *Dear Folks At Home* (New York: Houghton Mifflin Co., 1919), p. 16.

(27) National Archives Record Group 127, *op. cit.*

(28) *Springfield Sunday Republican*, February 1960.

(29) Interview of Miss Margaret L. Powers, p. 2, hereafter *Powers interview* (*Women Marines World War I File*).

(30) *Linscott interview*, p. 28.

(31) Interview of Mrs. Avadney (Hea) Cheney, p. 17, hereafter *Cheney interview* (*Women Marines World War I File*).

(32) Interview of Miss Ingrid Jonassen, p. 3 (*Women Marines World War I File*).

(33) *Powers interview*, p. 3.

(34) *Cheney interview*, p. 8.

(35) *Powers interview*, p. 2.

(36) *Ibid.*

(37) Jonassen interview, p. 3.

(38) Ibid., p. 2.

(39) Miller interview, p. 6.

(40) "Women in the Navy," Chapter XXXI, op. cit., p. 38.

(41) Marine Corps Orders No. 62 (Series 1918), HQMC, 7 December 1918.

(42) Short interview, p. 5.

(43) Springfield Sunday Republican, February 1960.

(44) National Archives Record Group No. 127, op. cit., memo from MajGen Barnett, dated 8 February 1919.

(45) Springfield Sunday Republican, February 1960.

(46) Interview of Mrs. Mabelle (Musser) Hall, p.2, hereafter Hall interview (Women Marines World War I File).

(47) Interview of Mrs. Lucy (Erwin) Winter, p. 1, hereafter Winter interview (Women Marines World War I File)

(48) "Women in the Navy," Chapter XXXI, op. cit., p. 39.

(49) Marine Corps Pay Tables in Register of the Commissioned and Warrant Officers of the U.S. Navy, U.S. Naval Reserve Force and Marine Corps (Washington, D.C., 1919).

(50) Jonassen interview, p. 3.

(51) Short interview, p. 8.

(52) Lawrence interview, p. 2.

(53) U.S. Marine Corps Muster Rolls for January 1919.

(54) Interview of Mrs. Ether (Hedtler) Budge, p. 2., hereafter called Budge interview (Women Marines World War I File).

(55) Ibid., p. 4.

(56) Ibid.

(57) Individual biographies (Women Marines World War I File).

(58) Interview of Miss Elizabeth Bertram, p. 2. (Women Marines World War I File).

(59) Ibid.

(60) Recruiters' Bulletin, Vol 5, No. 9 (October 1919), p. 5.

(61) Interview of Mrs. Edith (Macias) Vann, p. 2., hereafter Vann interview (Women Marines World War I File).

(62) Ibid., p.3.

(63) Lawrence interview, p. 2.

(64) Miller interview, p. 5.

(65) Linscott interview., p. 18.

(66) Interview of Miss Sarah Jones, p. 6 (Women Marines World War I File).

(67) Cheney interview, p. 14.

(68) Short interview, p. 6.

(69) Springfield Daily Republican, February 1960.

(70) National Archives Record Group No. 127, op. cit., Letter from MajGen Barnett, dated 8 March 1919.

(71) Linscott interview, p. 39.

(72) Miller interview, p. 5.

(73) Ibid., p. 7.

(74) Jonassen interview., p. 3.

(75) Linscott interview., p. 25.

(76) Leatherneck, 21 November 1918.

(77) Ibid.

(78) Miller interview., op. cit., p. 7.

(79) Leatherneck, 21 November 1918.

(80) Interview of Mrs. Charlotte (Titsworth) Austin, p. 3 (Women Marines World War I File).

(81) Leatherneck, 9 January 1919 and 20 February 1920.

(82) National Archives Record Group No. 127, op. cit., memo from MajGen Barnett, dated 15 July 1919.

(83) Linscott interview, p, 42; Springfield Sunday Republican, February 1960.

(84) "Women in the Navy," Chapter XXXI, op. cit., p. 38.

(85) National Archives Record Group no. 127, op. cit., letter from MajGen Barnett to Miss Sawyer, dated 30 June 1919.

(86) Ibid.

(87) Short interview, p. 8.

(88) The New Republic, Vol XVI, No 197 (10 August 1918), p. 33.

Appendix B

Bibliography

Magazines

Leatherneck. Vol. 26, No. 2 (February 1943).

Marine Corps Gazette, Vol. 20, No. 2 (May 1963).

The Marines Magazine. Vol. 3, No. 10 (September 1918); Vol 3, No. 11 (October 1918); Vol. 4, No.1. (December 1918).

The Recruiters' Bulletin. Vol. 4, No. 11 (September 1918); Vol. 5, No. 3 (April 1919); Vol. 5, No. 4 (May 1919); Vol. 5, No. 6 (July 1919); Vol. 4, No. 11 (September 1918); Vol. 5, No. 9 (October 1919).

Newspapers

Evening Journal. (New York), 29 January 1919.

Evening Mail. (New York), 19 August 1918, 29 January 1919.

Evening Star. (Washington), 7 August 1972.

Leatherneck. (Quantico), 14 November 1918, 21 November 1918, 9 January 1919, 16 January 1919, 6 February 1919, 13 February 1919, 21 February 1919, 19 June 1919, 20 February 1920.

New York Evening World. 23 January 1919.

Washington Post. 17 July 1918, 18 August 1918, 22 September 1918.

Other Sources

National Archives: Record Group 80 - Records of the Secretary of the Navy, General Correspondence 28553-510

Record Group 127 - Records of the United States Marine Corps (1913-38) 1535-55

Headquarters, U.S. Marine Corps, History and Museums Division, Reference Section. Subject file on Women Marines World War I.

Interviews

Personal interviews were conducted with the following named persons, copies of which are contained in the files of the Reference Section, History and Museums Division, Headquarters, U.S. Marine Corps.

BENSON, Mary A. (Mrs. Albert Eldred) April 1972.

BERTRAM, Elizabeth (Mrs.) July 1972.

BOND, Olive (Mrs. Miller), August 1972.

CHAPMAN, Mary Lou (Mrs. Warren Clayson), April 1972.

ELLYSON, Samia (Mrs. Pope), April 1972.

ERVIN, Lucy (Mrs. Winter), November 1972.

FRANTZ, Frieda (Mrs. T. Kelleher), October 1972.

GARDINER, Marie (Mrs. Buttell), October 1972.

GERTLER, Florence (Mrs. Albert Miller), November 1972.

GILL, Kathryn (Mrs. Harry Belden), December 1970.

HEA, Avadney (Mrs. Paul Cheney), August 1972.

HEDTLER, Louise, Mrs. Norman Budge), October 1972.

HOPKINS, Leola Anne (Mrs. Erenesto Beavers), August 1972.

JONASSEN, Ingrid, May 1972.

JONES, Sarah, November 1972.

KINNA, Pauline (Mrs. Chamberlin), October 1972.

KIRKHAM, Ida (Mrs. Herbert Lawerence), October 1972.

LAYCOCK, Fanny (Mrs. Karl Kadic), November 1972.

LEE, Edna (Mrs. Robert Freund), July 1972.

LEIBRAND, Lela (Mrs. Rogers), May 1972.

LINKINS, Carrie, October 1972.

MACIAS, Edith (Mrs. Vann), April 1972.

MCLUCKIE, Louise (Mrs. Davey), October 1972.

MUSSER, Maybelle (Mrs. William Hall), April 1972.

O'MALLEY, Lillian (Mrs. Joseph Daly), March 1972.

POWERS, Margaret, June 1972.

SHARKEY, Mary (Mrs. Joseph Short), July 1972.

SHOEMAKER, Elizabeth (Mrs. Linscott), August 1972.

SWOPE, Alma, October 1972.

TITSWORTH, Charlotte (Mrs. Austin), December 1972.

VAN WAGNER, Violet (Mrs. Lopez) February 1972.

WEIDINGER, Florence (Mrs. MacAlister), July 1972.

Appendix C

Partial Alphabetical Roster of Women Enrolled in the Marine Corps During World War I

ABERLE, Caroline O.

AHLUM, Frances C.

ALLEN, Susie

ANDREAS, Alfreda A.

ARNOLD, Jewell (Mrs. Craig Hansen)

ARNOLD, Judy

ARTHUR, Minnie

BAER, Ethel M.

BAILEY, Elsie M. (Mrs. Pinney)

BAKER, Emma L.

BALFOUR, Isabelle (Mrs. Morton)

BARKER, Alma

BARTLE, Maria

BARTON, Edith D.

BEALE, Rose

BECK, Maryann K.

BECK, May

BELLINGER, Virginia (Mrs. Bogart)

BENNETT, Mae E.

BENSON, Mary (Mrs. Albert W. Eldred)

BEPPLER, Elsie M. (Mrs. Schaeffner)

BERK, Alma (Mrs. Swope)

BERTRAM, Elizabeth

BITTNER, Elva D.

BLISS, Mary Ann (Mrs. Kennedy)

BOND, Olive M. (Mrs. Miller)

BOONE, Irene (Mrs.)

BOWMAN, Ann

BOYD, Justina M.

BRESNAN, Mary (Mrs. Edward Duggan)

BROADWATER, Alice

BROWN, Anna (Mrs. T. Stanley Holland)

BROWN, Eleanor F.

BUCKLEY, Sarah (Mrs. Daniel)

BUDD, Graham

BURGESS, Eleanor

BURNS, Mae (Mrs. Walter Fisher)

BYRD, Norma (Mrs. Watson)

CANTER, Lillian E.

CAREY, Emily J.

CAYIA, Jennie

CHAMBERLAIN, Lavinia P. (Mrs. Roy Kinna)

CHANDLEY, Pearl (Mrs. Howard L. Oagley)

CHAPMAN, Mayr Lou (Mrs. Warren Clayson)

CLARK, Grace

CLARKE, Blanche D.

CLAY, Mary

CLYNE, Rose M.

COLCLAZER, Alma (Mrs. Crosby)

COVERT, Marie (Mrs. Anders)

COWELL, Mildred A. (Mrs. Jones)

COXE, Elizabeth B. (Mrs. MacIntosh)

CROOP, Susan M.

CROSS, Helen M.

CUTTER, Lillian

DAOUD, Nahiba

DAVEY, Louise A. (Mrs. McLuckie)

DAVIDSON, Eva (Mrs. William Rabill)

DAVIS, Sybil B. (Mrs. MacNamara)

DEUTERMAN, Louise (Mrs. Allison)

DEVEREAUX, Margaret M. (Mrs. Jeschke)

DICKERSON, Bessie R. (Mrs. Carl G. Carlson)

DONAGHY, Clare M.

DOYLE, Mary B.

DRINKWATER, Charlotte M.

DUCKMAN

DUNN, Helen H. (Mrs.)

DUPONT, Helen C. (Mrs.)

ELLYSON, Jennie (Mrs. Gus Creange)

ELLYSON, Samia (Mrs. Pope)

ELSESSER, Caroline C.

ENGLISH, May

ERERLE, Carolyn

ERVIN, Lucy (Mrs. Winter)

EVANS, Matilda M. (Mrs.)

FALLON, Mary M.

FAUNCE, Elizabeth A.

FELL, Mary H.

FOLKS, Maybelle La Von (Mrs. Thad Taylor)

FOWLER, Beverly

FRANTZ, Frieda (Mrs. T.C. Kelleher)

FRENCH, Agatha G.

GABY, Minette (Mrs. Benjamin Duckman)

GAMAGE, Philena (Mrs. Reginald H. Beale)

GARDINER, Marie A. (Mrs. Buttell)

GARDNER, May

GASKINS, Mabel E.

GERNSBACHER, Viola

GERTLER, Florence (Mrs. Albert Miller)

GILL, Kathryn (Mrs. Harry J. Belden)

GRAHAM Frances C.

GRAHAM Ruth C. (Mrs. Ralph H. Crandall)

GRIFFITH, Virginia L.

GUNN, Elizabeth

HAINES, Ruth

HANSON, Genevive (Mrs. Arnold)

HARGIS, Mae E.

HART, Margaret (Mrs. Canby)

HASSON, Carol (Mrs. Miller)

HATSEL, Margaret A.

HAY, Mary (Mrs. Douglas)

HEA, Avadney (Mrs. Paul E. Cheney)

HEDTLER, Ethel Louise (Mrs. Norman D. Budge)

HELMES, Emma C.

HODGKINS, Alice W.

HOPKINS, Leola A. (Mrs. Erenesto Beavers)

HOYLE, Mary T.

HURLEY, Blanche

HUTCHINS, Irene (Mrs. D. A. Knight)

IRVINE, Blanche P.

JACOBSON, Myrtle (Mrs. H. W. Gunther)

JOHNSON, Elizabeth N.

JOHNSON, Opha M. (Mrs.)

JONASSEN, Ingrid

JONES, Norma (Mrs. Watson)

JONES, Sarah L.

KEARNEY, Jessie E. (Mrs. Arthur Perkins)

KELLER, Esther M.

KELLY, Mary C. (Mrs. Howard)

KENNY, Carrie E. (Mrs. Thompson)

KINNA, Lavina P. (Mrs. Chamberlin)

KIRKHAM, Ida G. (Mrs. Herbert F. Lawrence)

KIRKHAM, Patsy (Mrs. Owen)

KIRKPATRICK, Pat

KRAUSS, Margaret C. (Mrs. Noyes)

KUGAN, Janet C.

KURGAN, Wanda

LAKE, Theresa

LAMMERS, Sophia J.

LASALLE, Lillian R.

LAYCOCK, Fanny E. (Mrs. Karl Kadic)

LEE, Edna E. (Mrs. Robert F. Freund)

LEIBRAND, Lela E. (Mrs. Rogers)

LEWELLEN, Ethel (Mrs. J.B. Dennis)

LINGLE, Daisy (Mrs. Charles Myrtle)

LINKINS, Carrie

LOVELEY, Ethel R. (Mrs. Miloy)

LOWEN, Bertrude T. (Mrs. Toogood)

LYNCH, Maefaret

MABLEY, Mildred (Mrs. E. M. Donnelly)

MACIAS, Edith A. (Mrs. Vann)

MACIAS, Sara J. (Mrs. Brewer Stouffer)

MADLAND, Helen (Mrs. Orra Armstrong)

MAHAN, Alice L.

MALONE May F.

MARCUS, Rose (Mrs. Benjamin Hurvitz)

MARINA, Maria (Mrs. G. R. Kerm)

MASSENGALE, Mabel

MCGOLDRICK, Margaret C.

MCMORRIS, Victoria M. (Mrs. Sheevers)

MIDDLETON, Marjorie

MILLER, Mildred A. (Mrs. Richard Romer)

MILLINGTON, Anna L.

MILLINGTON, Clara

MILLS, Gladys E.

MILLER, Olive Mae (Mrs. Michael)

MONTGOMERY, Edith C.

MORINA, Maria L. (Mrs. G. R. Kern)

MORRIS, Anna (Mrs. McGoldrick)

MORRISON, Maud S.

MUIR, Charlotte E. (Mrs. Martin)

MULL, Helen M. (Mrs. Miller)

MUSSER, Mabelle L. (Mrs. A. William Hall)

NELSON, Elizabeth (Mrs. George T. Welk)

NEUMARK, Helen

NUBSON, Troy A.

O'KEEFE, May L. (Mrs. Hurley)

OKEY, Ouida M. (Mrs. William J. Sharp)

OLSON, Alma E. (Mrs. Gaston Stickler)

O'MALLEY, Lillian C. (Mrs. John Daly)

OMAN, Alma J. (Mrs.)

O'NEILL, Helen G.

OWEN, Maybelle (Mrs.)

PATTERSON, Lillian

PEZZULI, Helen

PIDGEON, Katherine N.

PINKSTON, Caroline A.

PINNEY, Elsie

POWERS, Margaret L.

PRIESS, Helen I.

PURVIS, Harriet R. (Mrs. Larsen)

QUINN, Doris B. (Mrs. Lemly)

REILLY, Mary (Mrs. Rasener)

REINHARDT, Doris B.

REISS, Mae C.

RIENKENS, Charlotte S.

ROBINSON, Dorothy L.

ROEDER, Helen A.

ROTT, Lynn R.

ROSE

ROUMAGE, Jeanette C. (Mrs. Kelsy)

ROUNDTREE, Anna C.

RUDDER, Willie C.

RYAN, Anna N.

SANDS, Alice

SAWYER, Lillian M.

SAWYER Ray C.

SAYLOR, Myrtle R.

SCALES, Grace E. (Mrs. Tippett)

SCHLIGHT, Marie S. (Mrs. Kane)

SCHNADER, Madeline

SCHROEDER, Bessie L. (Mrs. Patchen)

SCHROEDER, Irene

SETTLE, Sarah G.

SHARKEY, Mary (Mrs. Joseph Short)

SHAUGHNESSY, Margaret C.

SHELDON, Anna A. (Mrs. Fulkerson)

SHOEMAKER, Elizabeth (Mrs. Robert Linscott)

SISSON, Mildred G.

SLY, Ethel W.

SMITH, Bessie (Mrs P. Frederick Dryer)

SMITH, Dorothy F.

SMITH, Thea A.

SOCKRIDER, Elsie L.

SORENSEN, Mattie L.

SPIKE, Ruth A. (Mrs. Frank Spike)

STIMSON, Hazel L. (Mrs. Fitzgerald)

STREEKS, Emma L.

STRUM, Ruth W. (Mrs. Errett)

SWENSON, Enga A.

THEK, Martrese (Mrs. Ferguson)

THOMAS, Gladys (Mrs. W. L. Allen)

THOMPSON, Anna E.

THOMSPON, Carrie E. (Mrs. Kenny)

TITSWORTH, Charlotte (Mrs. Austin)

TITUS, Daisy E. (Mrs. Spernoga)

TRACY, Ina C.

TODD, Elizabeth M.

TULLY, Irene (Mrs. Rothrock)

TYSON, Edith (Mrs. Tom Shepard)

VAN EDSINGNA, Jane (Mrs. Arthur Blakeney)

VAN WAGNER, Violet (Mrs. Jose Lopez)

WASHBURNE, Ethel

WATSON, Eunice C.

WEIDINGER, Florence (Mrs. MacAlister)

WELLS, Jean C. (Mrs.)

WERNER, Lillian D.

WHEATLEY, Henrietta A.

WHITESIDES, Mabel H.

WIELAND, Clara L.

WILKINS, Florence G.

WILKINS, Louise B.

WILLIAMS, Mozelle M. (Mrs.)

WILSCHINSKI, Martha

WITHERS, Margaret C.

WITTICOS, Margaret

WRIGHT, Minnie (Mrs. William Sayers)

Appendix D

Authorization to Enroll Women Marines

NAVY DEPARTMENT

WASHINGTON

August 8, 1918

To: Major General Commandant, U. S. Marine Corps

Subject: Enrollment of women in the Marine Corps Reserve for clerical duty.

Reference: Letter of Major General Commandant, USMC, dated August 2, 1918.

1. Referring to letter of the Major General Commandant U.S.M.C. as per above reference and in particular to the statement contained in the second paragraph thereof, that it is thought that about 40% of the work at the Headquarters, U.S. Marine Corps can be performed as well by women as by men: authority is granted to enroll women in the Marine Corps Reserve for clerical duty at Headquarters, U.S. Marine Corps, Washington, D.C., and at other Marine Corps offices in the United States where their services may be utilized to replace men who may be qualified for active field service with the understanding that such enrollment shall be gradual.

Josephus Daniels

Appendix E

Wilchinski and Leibrand Articles

Articles written by Sergeants Martha Wilchinski and Lela Leibrand concerning their experiences in World War I.

Article from The Recruiters' Bulletin, Vol. 4, No. 12 (October 1918), p. 16.

New York, September 23, 1918

Dear Bill:

Well, Bill, here I am again. I've been waiting for a chance to write you, but you'd never believe me if I told you I've been so busy sweeping floors and picking up cigarette butts and washing windows and everything, I haven't had a chance. That's a fine occupation for a Marine! Believe me, if I had known what I was enlisting for-! What'll I say to my grandchildren, Bill? When they ask me: "What did you do in the Great War, Grandma?" I'll have to say: "Washing windows on the second floor." That's a fine thing to have written on your tombstone, isn't it!

It's these everlasting inspections, Bill. There's only one thing the matter with them - they never come off! That is how I have it doped out, Bill. They want to save janitor's wages, so an order comes around that the general or somebody is coming for inspection. Then everybody gets busy and starts scrubbing and bumping into each other and stepping on each other's feet and everything. And just because I'm low in rank I have to do most of the work. The captain comes in and says: "Here's where you ought to shine!" "I am shining, Captain," I says; "I'm shining everything in sight." So he takes a chair and puts it on the table and stands on it and runs his finger along the ceiling. Then he shows it to me, and it has a speck of dust on it, and he looks at me like I broke his heart. So I climb up after him and I'm so nervous, I fall down and break the chair and scratch the table and dislocate my adenoids.

But that isn't all, Bill. I don't want to get court-martialed or anything by the general - not until pay-day, anyway - so I wash my face and go out and get a shine and everything. And the general never came that day at all, and the next day it rained and spoiled my shine and I had to get another one. I tell you, Bill, this life is wearing me out to a whisper. I'm getting so thin I had to take another hole in my wrist watch.

You remember I told you last time that a "movie" man took a picture of me being sworn in. He said it was going to show the week after. Well, Bill, I went to the movies steady every night for three weeks and spend $2.13 on myself and that fellow Pete, who lives next door to us; he's got a B classification - he'll be here when they go and be here when they come back. And all I got out of it was Spanish influenza and a pain in the ear on the side where Pete was snoring all the time, and I'm out $2.13, and all the lunch I can live on for a month now is a glass of milk and a short walk.

My cousin Sadie has been coming in and bragging for a week about her friend's being gassed and wounded and decorated and everything. I know her friend all right. The only way he'd ever get on the casualty list is by starving to death on account of sleeping all day and not having to get up for meals. And the only decorations he'd ever get are a couple of black eyes. Well, I was might pleased when I saw your name in the papers, Bill. That Sadie hasn't got a thing on me now.

I hope they didn't have to give you ether, Bill. I hear the fellows do terrible things when they come out of ether. I heard of a fellow who made love to the nurse. I bet he tried to hold her hand and everything. I get a lot of inside dope about the war, Bill; that's how I happened to hear about this. You want to be careful about those French girls, Bill. They're terribly rough. I hear you can't take a peaceful walk by yourself without having one of them drape herself around your aesophagus and getting closer to you than your landlady on payday. I'm sending you a book on biology. It's called "How to Tell the Wild Women." You ought never to go to Paris without it.

I have some very difficult situations to handle sometimes, Bill. For instance: The other day the lieutenant and I were waiting to go down in the elevator. Now, here's the question. If I am a lady and he's a gentleman, I go in first. If he's an officer and I'm a corporal, he goes in first. It all depends on how you look at it. I didn't know how he'd take it, so I thought I'd wait and see what he'd do. I guess he thought the same thing. So we both stood there eyeing each other up on the right oblique. Then he stepped forward and I stepped back. Then he stepped back and I stepped forward. Then we both stepped back. I was getting pretty dizzy by that time. I guess he was too. Then we both squeezed in at the same time. I guess that's what they mean by military tactics.

Well, I've got to stop now and start in picking up cigarette butts again. They're smoking them very short this year. I tell you, Bill, if I ever get out of this alive and have a home of my own, there'll be no cigarettes in my house. Not that I'm throwing out any hints. I thought you'd just like to know, that's all.

Your comrade in arms
Corporal Martha

MARTHA THE MARINE!

The Uniform is the Thing!

IT'S "US CORPORALS" NOW, MARTHA TELLS BILL

Wearing Chevrons Is Fine, but Drill and "Osculations" - - - - - Ain't.

October 17, 1918.

Dear Bill:—

Of course, Bill, you must have heard of my promotion. I'm a regular non-commissioned officer now with chevrons and a swell head and everything. When they call me Corporal, I tell you Bill it feels like Napoleon.

The Colonel came up for the occasion He called me out in front of everybody. "Private," he said, "Were you ever served with a warrant?" "No sir," I said, wondering what I had done now. I've managed to keep out of jail until now. "Well," he goes on, "you're going to be now." If he did have anything on me I thought it was kind of mean to call me down in front of everybody. "I [illegible] a wonder," I said, "you wouldn't shout it from the roof." But I said it to myself.

But when I read the warrant I saw that I had suddenly been changed into an officer. You know when a man is made a lieutenant the Major or somebody always pins on his shoulder bars, so I was waiting for him to sew my chevrons on. But he didn't. I guess he's not very handy that way.

There are a lot of things you have to go through in this life, Bill. The other day I had to go over to the doctor to be osculated or something. And what with getting vaccinated and feeling like the measles and cutting a back tooth at the same time, I felt as if all I needed was a hoop and a nursemaid to make my happiness complete.

I'll tell you about the scientific explanation of this osculation business, Bill. The doctor takes a tool about the size of a bicycle pump, jabs it into your arm until it comes out of the back of your neck and begins pumping typhoid and yellow fever germs into you at the rate of a million a second. By the time he gets through with you, you feel like the cross section of an ant hill. Now these here germs are supposed to be dead but there's still a lot of fight in them. And as soon as they get inside of you they suddenly remember all the family feuds they had in their last residence and begin to revive them. Then the germs that were there before they came get sore and start a free for all fight and that sets up what they call a typhoid immunity. Which means that if a single typhoid germ ever stays far enough from home to get into you he'll soon wish he'd followed father's advice and stayed at home. I hope you understand this explanation Bill, because then maybe you can explain it to me.

All the girl Marines drill on the roof every day.

You never can tell about those drill commands. For instance. The drill sergeant says "fall out!" Now there's a little blonde not a thousand miles away from here with whom I've been wanting to have a falling out for a long time. So when the sergeant gives me this invitation I don't wait to be told twice but go right to it. They tried to stop us but it was too late. The damage done was terrible.

Then this right about face business. You never saw such a mess. At the Command "about" you drape your right leg about your left, the feet toeing in, the knees concentric and two feet apart; at the command "face" you swing about, your fist resting in your neighbor's right eye while you land gracefully on his feet. Now anybody can see that a stunt like that is bound to end up in complications. Why you can start a war in a minute just by standing six men up and telling them to about face into each other.

I've been looking six ways for my uniform for weeks, until an order comes out that you have to send your measurements to Philadelphia or Jersey City or some place and have your uniform made by mail. Now I've had enough experience with these correspondence schools to know better. Don't you remember when my father tried to learn the cornet in ten lessons by mail guaranteed and the neighbors complained so much that we had to move out? Well I expected the worst and I can't say I was much disappointed. When I put my uniform on I looked like a physical map of Colorado.

I'm learning the duties of an officer very fast, Bill. I've got up to chewing tobacco. I start on that next week. I can't swear yet but I know all the words. We'll be able to understand each other better when you get back Bill.

Good-bye Bill from
Friend Corp.

The Girl Marines

By Pvt. Lela Leibrand

THREE OF OUR GIRL MARINES
Here's the way the Heavy Artillery Force at Headquarters looks in full uniform. Left to right: "Big Bertha" Macias, "Minnenwerfer" Bennett and "Flammenwerfer" Macias

Cherchez la femme! (Find the woman!) It is no longer a problem down at Headquarters in Washington. Girls, girls everywhere! And the Marines might just as well accustom themselves to us for we've come down among them to stay four years!

The moment your Marine Corps sent out the call for girls we flocked to the recruiting stations in every village, hamlet and town, eager to be one of that splendid body of men who have rendered such excellent account of themselves "overthere." And, we found they were most particular who they enrolled. "One hundred per cent men; one hundred per cent women," seemed to be the slogan, resulting in about four hundred being chosen out of as many thousand applicants. Believe us, we who are in are mighty glad it was us! It's an enormous satisfaction to know you can rate such an organization.

Many of us have left splendid positions back home to answer the call of our country in her hour of stress, even as the boys, because we knew we could be useful. Others of us have come because the Marine Corps pays a better wage than most private concerns for the same kind of work, and the hours, 8.30 A.M., to 4.30 P.M., are especially light. Then, not the least inducement is that fascinating uniform,—enough to make any girl leave home. Have you seen them? Done up in one we have to salute officers and everything.

You see, we are here to attend to your official business while you go out to fight. Some of the officers and men are most skeptical as to our intentions and we've found we must convince them we mean business. (Mr. Stroschein please note.) We are going to do it too, by honest effort, rigidly applied. Our work is exacting, and the detail cruel, in a way, for it keeps us upon the tips of our mental toes every moment of each day. To trifle means an error; errors mean inaccurate records. It can't be done in the Marines! However, the men are kind and considerate and patient with us, though a bit dazed at the newness of it.

There's romance in the work; real, live, human romance. We keep in closer touch with you than you dream. If you are transferred we know where from and to. If you are hurt, we know when, where and how much. When you reach another step in progress of your training, marksmanship, promotion, we hear about it and see to it you get the proper credits on your records and—the additional pay! But, when we read "Ashore France, servg. with Army fr," stamped on your card—well, it is but a little hope, just a thought, but it is a prayer that God will watch over you and care for you, in the end delivering you safely home, truly glorified, to that little mother who writes us occasionally that we might know exactly where she may be reached the most speedily "in case of emergency." Romance, did you ask? Each one of you is a special little story all by yourself. Sssh! We know when you're brigged, too!

Oh, I mustn't forget to tell you about our drills. You boys would turn dark green with envy to see us "right face," "left face," salute" (at the heathenish hour of 8.30 in the morning). And, Corporal Lockout, who is in charge of our drills said (and these are his exact words), "Girls learn the drills much easier than men." He added he doesn't know why, but we do. Besides Corporal in Charge Edward E. Lockout, we have Sgt. Arthur G. Hamilton, Corpl. Guy C. Williams and Pvt. Herbert S. Fitzgerald, each in charge of a company and each one enthusiastic about it. Well, who wouldn't be?—Am I right? We do police duty in our offices, too, but we didn't have to be drilled to that.

However, Corporal Lockout whispers to me that the drilling is only a means to an end. Discipline is the goal! Military discipline, at that. Personalities and sex must be subdued. The girls must learn as the men have learned, they are privates in the Marine Corps; must learn to accept discipline as the men accept it, without a single consideration for the fact that in private life they were self-governing young ladies. There is to be no proviso to this discipline. We are Marines! That says it all! And, once our lesson has been told to us, I hear from various sources, in ominous tones, leniency will cease to exist. The chaff will be separated from the wheat. A sort of forewarning. Now, do you gather a bit of our importance among you? We are not a fad by any means.

The Navy Department has taken over a hotel for us the Hotel Vendome, at Third and Pennsylvania Avenue, and every convenience and entertainmet of home will be supplied us there at a charge much within the reach of our purses. Too, the boys at Quantico ask us down, a few at a time, to their dances, and we are chaperoned by Colonel McLemore and his wife and daughter. We haven't a moment to get lonely or homesick.

All this just to let you know we are here, also to warn you that, whatever you do, remember the eagle eyes of the Marinettes are right upon you. Watch your step!

And just so you'll know who we are, here's the list of all the Marinettes now on duty at Headquarters, excepting the girls whose names were published in the October issue:

ADJUTANT AND INSPECTOR'S DEPARTMENT

Privates

Mary C. Sharkey, Philadelphia, Pa.
Mary H. Fell, Elmira, N. Y.
Ida G. Kirkham, New York
Mildred A. Miller, Hoboken, N. J.
May L. O'Keefe, Jersey City, N. J.
Dorothy F. Smith, New York
Ruth A. Spike, Brooklyn, N. Y.
Mary A. Bresnan, No. Cambridge, Mass.
Pearl Chandley, Inglefield, Ind.
Bessie R. Dickerson, Shelbina, Mo.
Philena E. Gamage, Waltham, Mass.
Ethel L. Hedtler, Mattapan, Mass.
Mary T. Hoye, Brookline, Mass.
Margaret L. Powers, Quincy, Mass.
Rose Marcus, Boston, Mass.
Daisy E. Titus, Moline, Ill.
Frances C. Ahlum, Osage, N. J.
Ruth C. Graham, New York, N. Y.
Sarah L. Jones, Meridien, Mass.
Anna E. Morris, West Orange, N. J.
Helen H. Neumark, New York, N. Y.
Anna C. Rountree, Jersey City, N. J.
Jennie F. Van Edsinga, New York
Ethel W. Sly, Nucla, Colo.
Mary M. Beck, New York, N. Y.
Elsie Beppler, New York, N. Y.
Mary Fallon, New York, N. Y.
Mary Doyle, New York, N. Y.
Helen Dunn, New York, N. Y.
Ruth Haines, Washington, D. C.
Mary A. Williams, Youngstown, Ohio.
Helen M. Cross, Baltimore, Md.
Mildred A. Cowell, Marietta, Ohio
Ouida M. Okey, Woodsfield, Ohio
Ethel R. Loveley, Washington, D. C.
Mary Ann K. Beck, Rochester, N. Y.
Evadne B. Hea, Buffalo, N. Y.
Florence G. Wilkins, Washington, D. C.
Eleanor F. Brown, La Plata, Md.
Anna H. Brown, La Plata, Md.
Clare M. Donaghy, Oil City, Pa.
Willie C. Rudder, Washington, D. C.
Blanche P. Irvine, Washington, D. C.
Maybelle L. Folks, Clarksburg, Ind.
Margaret C. Withers, Yonkers, N. Y.
Dorothy L. Robinson, Baltimore, Md.
Grace Clark, New York, N. Y.
Margaret M. Devereux, Washington, D. C.
Genevieve H. Hansen, New York, N. Y.
Sylvia L. Reinhardt, Brooklyn, N. Y.
Lela E. Leibrand, New York, N. Y.
Mary C. Titsworth, Washington, D. C.
Lillian C. O'Malley, Washington, D. C.
Mildred G. Sisson, Washington, D. C.
Margaret A. Hatzel, Rockville Center, L. I., N. Y.

QUARTERMASTER'S DEPARTMENT

Privates

Irene Hutchens, Allegan Co., Mich.
Gladys E. Mills, Washington, D. C.
Elizabeth A. Faunce, Washington, D. C.
Edna E. Lee, Olivet, Mich.
Elsie M. Pinney, Baltimore, Md.
Gladys Thomas, Greeley, Iowa.
Minnie Wright, Preston, England.
Ethel Washbourne, Rock Falls, Ill.
Virginia Griffith, New York, N. Y.
Carolyn Elsesser, Williamsport, Pa.
Gertrude Lowen, Bournemouth, England
Clara Wieland, Brooklyn, N. Y.
Edith Tyson, Crab Orchard, Neb.
Carel Hasson, Minneapolis, Minn.
Hazel Stimson, Detroit, Mich.
Carrie Linkins, Washington, D. C.
Blanche Clarke, Virginia, Ill.
Bessie Smith, Prince Georges Co., Md.
Mabelle Musser, Hinckley, Ohio
Harriet Purvis, Pittsburgh, Pa.
Lavina Chamberlaine, Washington, D. C.
Mabel Gaskins, Lancaster, Pa.
Elizabeth Gunn, Greenfield, Ohio
Anna Sheldon, Washington, D. C.
Jean Wells, Fullerton, Cal.
Clara Millington, Milwaukee, Wis.
Sybil Davis, Berwyn, Md.
Helen Roeder, Upper Marlboro, Md.
Susie Allen, Alexandria, Va.
May Malone, Washington, D. C.

PAYMASTER'S DEPARTMENT

Privates

Irene E. H. Bonne, Tuscaloosa, Ala.
Sara V. Buckley, Washington, D. C.
Lillian E. Canter, Plainfield, N. J.
Nahiba Daoud, Washington, D. C.
Eva Davidson, Harrisburg, Pa.
Charlotte M. Drinkwater, Washington, D. C.
Margaret C. Krauss, Washington, D. C.
Wanda Kurgan, New York, N. Y.
Fanny E. Laycock, Leesburg, Va.
Emma L. Streeks, Washington, D. C.

BANKRUPTCY NOTICE

Selling out cheap. Good collection of crowns, thrones, swords, robes, uniforms, panoply and other whatnots of royalty. Would make good collection for enterprising motion picture company. Communicate with

GERMANY, AUSTRIA, BULGARIA, et al,

National Agents.

Article From LEATHERNECK (Quantico) Vol. 2, No. 9 (2 January 1919).

FAIR MARINE TELLS OF FLIGHT IN HYDROPLANE

By Pvt. Lela Leibrand

Did you see the 894 circling about over your heads the other day, dipping, vooming, tipping and twirling? Well--that was I'M. How did I look?

You know,people have told me time and again the terrible experiences they had in airplanes; funny things happening to their stomachs when they were descending and twirling, and such and now I don't believe any one of those people who told me those things ever went up in an airplane, hydroplane or Marine plane. Honestly, it's just the opposite. You want to sing and laugh and dance. You leave the world behind and its troubles and worries and cares, and you just sail up among the clouds getting closer and closer to heaven and the quiet and peace you've always been told heaven means. You cannot remember a single unpleasant thing people have done to you or that you would like to do to people. You just sit there and drink in the air and look and look and look.

Try to Frighten Her

Captain Mims and Captain Page, our two well-known flyers, tried their best to frighten me before we started. Captain Page tied my helmet under my chin, and as he did so, told me in all seriousness that I must hold my mouth tightly closed to insure the imprisonment of my heart as the plane would leave the water. Captain Mims, drawing on those snappy looking overalls of leather he wears when flying, and incidently looks so ducedly handsome in, listed all the different antics he would perform once he got me into the air. But, I did some quick thinking, and just smiled back at them, for wherever Captain Mims sent me he had to go himself, so, I'd take a chance.

They all seemed to be somewhat worried because I showed no signs of weakening, and I suppose the crew thought I was one of those girls who remain seemingly passive throughout the preliminaries, then go all to pieces at the real test. But I showed "em, I did!

After I was strapped into my pit (Yes, hydroplanes have them); I became a little impatient. How was I to know it took hydroplane chauffeurs so long to get their machines started? Now, my Packard--yes, and in the movies, the driver just

Private Lela Leibrand stands on the wings of a hydro-
plane at Quantico after taking her first flight.
(USMC Photo 518553).

steps blithely into his seat and moves away from the earth. But, Captain Mims speeded the engine, then stopped it, then speeded it again. Poor me, I thought something had happened and it wouldn't go and I wasn't going to get my ride. Then, all of a sudden, we were off. I didn't realize we had left the water until I saw it receding below me. Then up, up, up.

O, Boy! There's not a single adjective in the United States that has the power to express it. I wanted to sing, so I burst forth, that is, I'm sure I did, but you know those engines! I know I never sang better in my life, either, for I was truely inspired.

Says She Wasn't Dizzy

Afraid? Why there is absolutely nothing to make one feel insecure up there. You are in that enormous, luxuriant plane and you are far too high to get dizzy looking down.

If you remember, that day was one of those clear, opal-pink days with just a hint of fog in the air to tone down the coloring. After we got a thousand feet or so above the friendly old Potomac, all sound and motion other than our own, ceased. I was looking, just filling my eyes full, and I've never seen a more beautiful sight, a more inspiring sight in my life.

Do you remember when you used to look at those stereopticon views grandmother used to have on the front parlor table? Looking through those goggles from that distance, took me back to those days, for that is just the way the earth looked like to me. Just a beautiful old painting by a master of color.

We circled about for a while, then--a bolt from the blue! Those standing on the shore awaiting our return heard a distant but terrible roar like an echo from Belgium or Alsace-Lorraine! There was a splash and a geyser of water from the heart of the old Potamac directly beneath the position of our plane. Captain Mims saw it and changed his course. Indian Head were proving out one of their big guns, but they shouldn't prove on us. It wasn't long before we came down. You see, being a green flyer I missed all this, and I tried to tell Captain Mims I didn't want to go down, but he couldn't hear me. He knew I wasn't safe up there so he was getting me where I would be safe.

We came down and we had chow and I saw the show with the other Marines, and then came home on a late train, tired but happy. I'm coming down again soon.

Made in United States
Cleveland, OH
06 November 2026

47678357R10050